A World View of Environmental Issues

second edition

To accompany selected texts from
Saunders College Publishing physical science series.

Leroy W. Dubeck
Frederick B. Higgins
Robert Patterson
Rose Tatlow
Catherine Ward
Barbara Wright

Saunders College Publishing
Harcourt Brace College Publishers

Fort Worth Philadelphia San Diego New York Orlando Austin
San Antonio Toronto Montreal London Sydney Tokyo

Copyright © 1998, 1995 by Harcourt Brace & Company

All rights reserved. No part of this publication may be reproduced or transmitted in any form or by any means, electronic or mechanical, including photocopy, recording, or any information storage and retrieval system, without permission in writing from the publisher.

Requests for permission to make copies of any part of the work should be mailed to: Permissions Department, Harcourt Brace & Company, 6277 Sea Harbor Drive, Orlando, Florida 32887-6777.

Printed in the United States of America

Kirkpatrick: <u>A World View of Environment Issues to accompany *Physics: A World View* Third Edition.</u> Dubeck.

0-03-024393-9

7 202 765432

ABOUT THE AUTHORS

Dr. Leroy W. Dubeck is Professor of Physics at Temple University.

Dr. Frederick B. Higgins is Professor of Environmental Engineering and Director of the Environmental Center at Temple University.

Dr. Robert Patterson is Professor of Environmental Engineering at Temple University.

Ms. Rose Tatlow is a media consultant who has collaborated in the writing of science textbooks.

Ms. Catherine Ward, Esq. is an environmental attorney practicing in New Jersey and Pennsylvania.

Ms. Barbara Wright is Reference Librarian in the Reference and Information Services Department of Temple University's Paley Library.

ACKNOWLEDGMENTS

The authors wish to thank Jennifer Steinberg for the computerized layout of the final copy, Robert Singer for drawing all of the illustrations, and Marc Sherman, Senior Associate Editor of Saunders College Publishing, for guiding us through the many stages of development and production.

This material is based, in part, upon work supported by the National Science Foundation and Grants No. DUE-9354383 and No. DUE-9455042. Any opinions, findings, conclusions, or recommendations expressed in this publication are those of the authors and do not necessarily reflect the views of the National Science Foundation.

In addition, this material is based, in part, upon work supported under a minigrant from the American Bar Association for Justice and Education. Any opinions, findings, conclusions, or recommendations expressed in this publication are those of the authors and do not necessarily reflect the views of the American Bar Association.

Table of Contents

CHAPTER 1: INTRODUCTION 1
    Science and Technology 1
    The Scientific Method 1
    Units and Scientific Notation 2
    Risk Assessment 2
    How to Research and Update Your Knowledge of an Environmental Issue .. 2
    Format for Chapters 3 - 6 3
    Technological Developments and the Legal System 3
    Sources of American Law 3
    Concepts Important to Environmental Law 4
    Science, Technology, and the Media 6
  Bibliography 6

CHAPTER 2: HOW TO RESEARCH ENVIRONMENTAL ISSUES:
        SCIENCE RESEARCH FOR NON-SCIENTISTS 7
  Where to Start 7
  Background Information 7
  Depth vs. Timeliness: Books 8
  Current Information: Magazine/Journal/Periodical Articles 8
  Yesterday's Headlines and Last Week's Magazines 11
  The "Information Superhighway": the Internet 12
  Legislative Update 12
  The Frustration Factor 13

CHAPTER 3: NUCLEAR POWER PLANTS, NUCLEAR WASTE DISPOSAL,
      AND OTHER RELATED ISSUES 14
    The Basic Physics of Nuclear Power 14
    Conventional Nuclear Reactors 15
    Breeder Nuclear Reactors 16
    Safety Issues Involved in Operating Conventional Nuclear Reactors 16
    The Three Mile Island Accident 16
    Safety Improvements in Reactor Designs 18
    Nuclear Plant Efficiency 18
    The Chernobyl Accident 19
    The Status of the Nuclear Industry in the Former Soviet Union 20
    Decommissioning of Nuclear Reactors 21
    Disposal of Radioactive Wastes 22
    Nuclear Weapons 23
    The Effects of Radiation On Humans 26
    Typical and Maximum Permissible Radiation Doses 27
    Radiation Experiments Conducted On Humans 27
    Operation Desert Storm and Depleted Uranium Weapons 28
    Irradiated Food 28
  Society's Reactions to These Environmental Issues 28
    History of Regulation 28
    The Handling, Transporting, and Disposing of Nuclear Materials 29
  Relevant Educational Audio-Visual Materials 30
  Internet Resources 31
  Bibliography 31
  Discussion Questions 32
  Notes 33

CHAPTER 4: INDOOR AIR POLLUTION 35
    Radon 35
    Smoking 40
    Asbestos 40
    Other Harmful Substances 41
    Sick Building Syndrome 42
    Indoor Air Pollution Controls 43
  Society's Reactions to These Environmental Issues 44
    Application of Existing Laws to the Indoor Air Pollution Issues 44
    Common Law and Other Approaches 45
    Future Controls 46
  Relevant Educational Audio-Visual Materials 46
  Internet Resources 46
  Bibliography 48
  Discussion Questions 48
  Notes 49

CHAPTER 5: THE GREENHOUSE EFFECT AND STRATOSPHERIC OZONE DEPLETION 50
    The Greenhouse Effect 51
    Sources of Greenhouse Gases 51
    Effects of Global Warming 51
    IPPC Report on Global Warming 53
    Methods of Diminishing the Effects of Greenhouse Gases 54
    What Can You Do to Prevent Global Warming? 54
    Ozone Layer Depletion 55
    Effects of Ozone Layer Depletion 57
    Some Myths about Stratospheric Ozone Depletion 58
    Methods of Diminishing the Use of Ozone-Destroying Chemicals 58
    What Can You Do to Prevent Stratospheric Ozone Depletion? 59
  Society's Reactions to These Environmental Issues 59
    Control-Based Legislation 59
    Incentive-Based Legislation 61
    Process-Oriented Legislation 61
    Role of the States 62
  Relevant Educational Audio-Visual Materials 62
  Internet Resources 63
  Bibliography 63
  Discussion Questions 64
  Notes 65

CHAPTER 6: ELECTROMAGNETIC WAVES 66
    Electromagnetic Radiation 66
    60-Hz Electric and Magnetic Fields 67
    Field Sources and Levels 68
    Interaction With Humans 69
    Health Effects of Power-Frequency Fields 70
    Effects in Cells and Tissue: *In Vitro* Studies 70
    Effects in Laboratory Animals: *In Vivo* Studies 71
    Effects in Human Populations: Epidemiological Studies 71
    Exposure Guidelines 72
    Radio-Frequency and Microwave Radiation 72
    Ultraviolet Radiation 73
  Society's Reactions to These Environmental Issues 74
    History of Government Regulation of High Voltage Lines 74
    Society's Ability to Affect the Placement of High Voltage Lines . 75
    Condemnation Process 76
    How Biological Risk Posed by EMF Affects the Legal Process 76
  Relevant Educational Audio-Visual Materials 77
  Internet Resources 77
  Bibliography 78
  Discussion Questions 80
  Notes 81

CHAPTER 7: THE PORTRAYAL OF ENVIRONMENTAL ISSUES IN FILM AND IN NON-DOCUMENTARY TELEVISION PROGRAMS 82
- **Films Related to Chapter 3** 83
  - The China Syndrome 83
  - Silkwood 84
  - Them! 85
  - Fat Man and Little Boy 86
  - Chain Reaction 86
- **Films Related to Chapter 4** 87
  - The Incredible Shrinking Woman 87
  - Safe 87
- **Films And Television Programs Related to Chapter 5** 88
  - Star Trek: The Next Generation—"A Matter of Time" 88
  - Star Trek: The Next Generation—"When the Bough Breaks" 88
  - Highlander II: The Quickening 89
  - The Arrival 89
- **Television Program Related to Chapter 6** 90
  - Star Trek: The Next Generation—"The Enemy" 90

1
Introduction

This supplement is intended to illustrate, through discussion of several environmental issues, how the study of physics is relevant to important real world issues. To provide additional context for your studies, the authors have included discussions of how American society has attempted to address environmental issues through legislation. In addition the authors describe feature films and television programs which convey society's concerns about these issues. Documentary films and other audio-visual resources are included at the end of each chapter.

Science and Technology

Science is concerned with discovering relationships between observable phenomena and with organizing and describing these phenomena in terms of theories. Technology is concerned with the tools and techniques for putting to use the discoveries of science. An important difference between science and technology is their impact on human beings. Unlike science, advances in technology can be measured in terms of their direct impact on our lives. Furthermore, while an individual may refuse to believe in a scientific theory such as the planetary model of the atom, a theory cannot be ignored once technology is created based on that theory: we do not have the option of living in a world without nuclear power plants and nuclear weapons.

The Scientific Method

One also needs to understand the scientific method in order to be able to properly assess potential hazards associated with environmental issues. This method can be described as follows:

      (1) recognize that a scientific problem exists
      (2) make an educated guess about the solution to the problem, namely, state a hypothesis
      (3) predict the consequences if the hypothesis is true
      (4) perform experiments to see if the predictions occur
      (5) formulate the simplest general rule that organizes the hypothesis, prediction, and experimental results into a theory

A **scientific theory** is a synthesis of well-tested and verified hypotheses about some aspect of the world around us. When a scientific hypothesis has been confirmed repeatedly by experiment, it may become known as a **scientific law** or **scientific principle**. A **scientific fact** may be defined as an agreement by competent observers of a series of observations of the same phenomena. From time to time scientific facts are revised by additional data about the world around us. Scientists often employ a **model** in order to understand a particular set of phenomena. A model is a mental image of the phenomena using terms with which we are familiar. For example, there is the planetary model of the atom in which scientists visualize the atom as a nucleus with electrons revolving about it in a manner similar to the way planets revolve around the sun. While this model is useful in understanding the atom, it is an over-simplified description of a real atom and thus does not predict all of its attributes.

Units and Scientific Notation

Metric units will be used throughout this supplement. Distances are given in meters, masses in kilograms, and time in seconds. The energy is then in units of joules and the unit of power is the watt, which equals 1 joule/second. Frequencies are presented in units of Hz (hertz), kHz (kilohertz or 1000 Hz), MHz (megahertz or 1,000,000 Hz), and GHz (gigahertz or 1,000,000,000 Hz). Similarly, voltages are presented in units of volts and kV (kilovolts). Small units are represented by prefix m (milli = .001) or μ (micro = .000001).

Scientific notation is used throughout the supplement which means that large and small numbers are given as powers of 10. For example, 3,000,000,000 equals 3×10^9. The power of 10 (i.e. the raised number to the upper right of 10) gives you the number of zeroes following the non-zero digit. Large numbers having more than one digit (excluding the zeroes) can be treated as follows: start with the number as given. Place the decimal point after the first digit. Then count the number of places that the decimal point has been moved to the left. This gives you the exponent of 10. Note that one continues to write all of the non-zero numbers. You change only the position of the decimal point and replace zeroes by a power of 10. As further examples, 186,000 equals 1.86×10^5, and 314,000,000 equals 3.14×10^8.

Very small numbers can also be expressed as a power of 10. Here the decimal point moves to the right and is represented by a negative exponent of 10. For example, 0.00000635 equals 6.35×10^{-6} since the decimal point was moved 6 places to the right. Note that you continue to write all of the non-zero numbers.

Risk Assessment

There are risks associated with almost every activity in life. Even miracle drugs designed to save lives occasionally cause allergic reactions that are fatal to a patient. Similarly, the production and use of energy is never completely safe. It is important to scientifically and dispassionately evaluate the benefits versus the risks of the use of a given material, process, or energy source. Sometimes these are difficult to assess accurately because one is not yet able to fully quantify some of the risks. This supplement will make estimates of some risks relevant to the discussions.

How to Research and Update Your Knowledge of an Environmental Issue

Since the information about the environmental issues described in this supplement will become somewhat outdated with time, it is important to be able to quickly locate the most recent and reliable information about a given topic. Any textbook about the environment is out-of-date by the time it is published because new information is constantly becoming available. For example, the alleged dangers of exposure to electromagnetic waves have been reported to be either very serious or non-existent from one week to the next! The reported dangers of radon in the home have similarly varied dramatically from one research study to the next. Consequently, we have devoted Chapter 2 to describing how to use print and interactive on-line resources available in most college and university libraries for researching environmental issues.

Format for Chapters 3 - 6

Each chapter dealing with environmental issues will have the following components:

      (1) a description of the scientific and technological aspects of the environmental issue(s)
      (2) society's reaction to the environmental issues as evidenced by legislation and guidelines issued by regulatory agencies
      (3) a listing of relevant audio-visual materials
      (4) a listing of relevant Internet sites
      (5) a bibliography of relevant books and articles
      (6) questions for discussion, use on examinations, and/or use as term paper topics

Technological Developments and the Legal System

Virtually every act in our society has legal consequences. Our system of government and laws goes far beyond a simple civic lesson of how a bill is turned into law and actually establishes the framework by which we, as Americans, deal with issues that affect us as a community and as a country. Nowhere is this more apparent than in the field of environmental science. As a society, we deal with new issues and concerns by attempting to understand and re-define them according to our existing societal/legal structure. The assimilation of new or changing concepts and issues into American society depends heavily on how our government categorizes and addresses them and an issue cannot be said to truly be recognized by American society until the government or another arm of the legal system has taken action respective to it.

Through the legal processes of "legislation, regulation and litigation," American society "massages" a concept or issue or concern until it fits into the framework of our society. Unfortunately, when the issues are science-related, sometimes the rush to legislate forces inexact or incomplete science to be considered precise and finalized. This can create mistaken impressions and cause unnecessary problems for those who must comply with the new laws and regulations. For example, when the federal government promulgated regulations supporting the Resource Conservation and Recovery Act in 1980, the regulations contained detection limits for certain chemical compounds and elements that the instrumentation of the day could not reliably reach. The law spurred the development of newer, more sensitive instruments to meet the need defined in the regulations, but in the interim caused substantial compliance problems for the regulated industries.

In order to understand the United States' attempt to deal with the science underlying environmental issues such as nuclear energy, ozone depletion, greenhouse effect and indoor air pollution, a review of the history and sources of American law is helpful.

Sources of American Law
—The Common Law

The American system of law has its roots in the English common law system. In general, there are two types of legal systems: "common law" and "civil law." A "common law" system is based on statutes and the existence of courts which interpret the statutes' application to specific facts, and which also interpret the decisions of other courts on the same issue (called "precedent"). The second system, "civil law," represents many of the European systems of law. Under civil law, a court interprets a statute but is not bound to follow precedent. This leads to inconsistent rulings for similar fact situations and makes the outcome of a case difficult to predict.

The system of common law which we enjoy today began in England in the 13th century. This was a period of time during which the feudal system was beginning to weaken. The feudal system was a land-based system whereby the king granted power and land to lords throughout the country and each lord was responsible for keeping those peasants and serfs who farmed the lands in line and obedient to the king. During the 13th century, people began to congregate together and develop cities, causing the land-based economy and the power of the lords to wane. In order to keep order and ensure control, the king appointed judges in the cities to decide controversies among the people, in the king's name. As these courts were initiated, there was virtually no consistency among their decisions and very few decisions were written down or recorded.

Over time this changed and by the time the American colonies were ready to choose a system of law, the common law system which had evolved in England was a highly structured, tiered system of courts which consulted and followed precedent and all of whose decisions were recorded. For their new country, despite whatever other reservations the colonists may have had about England, American judges followed the common law tradition. While this was controversial, the colonies gradually accepted the practice, and to this day only one state does not follow the common law system: Louisiana remains a civil law state.

—*Statutes and Ordinances*
Statutes and ordinances represent formal acts of legislation by governing authorities. Statutes are the acts of state and federal governments while ordinances are the acts of municipal governments (towns and cities). These laws can be very specific or very broad, but serve to establish a foundation. Generally, statutes include a delegation of authority to a particular agency or other government division which has the expertise to develop regulations or rules which flesh out the intent of the statute. For example, while a statute may contain a specific goal (such as "minimizing water pollution"), it is the substance of the regulations which provides the specific guidance for achieving that goal (by setting contaminant limits, requiring discharge permits, etc.).

—*Rules and Regulations*
As discussed above, the legislating body, whether it be at the federal level or the individual state legislatures, delegates some of its law-making authority to other sections of the government which are charged with the responsibility of overseeing and carrying out the statutes. These agencies create and then promulgate, through publication and public comment process, rules and regulations which provide specific guidance for how the intent of a statute is to be accomplished.

Rules and regulations must meet two constitutional standards. The promulgating agency must demonstrate that the proper procedure was followed in the creation and adoption of the rule or regulation, a process which is referred to as "procedural" due process. Second, the agency must show that the proposed rule or regulation falls within the bounds of the intent of the legislature (both explicit and implicit) when it drafted the statute. This is referred to as "substantive" due process. Rules and regulations which fail to meet either standard are subject to attack by lawsuit and will fail to be binding. If the rule or regulation is both procedurally and substantively correct, the rule or regulation will have the same status as any piece of legislation.

—*Constitutions*
The federal Constitution and the various state constitutions represent another source of law. These documents remain subject to changing and sometimes conflicting interpretations. Constitutions provide valuable safeguards against government abuses by protecting individual rights, protecting against governmental takings of private property, and ensuring the legitimacy of the rule-making process.

Concepts Important to Environmental Law
Environmental issues do not exist in a legal vacuum. The principles which are relevant in environmental law cut across numerous legal disciplines and include non-legal principles as well. Certain legal concepts which may have originated in other areas of the law have special significance when impacted by environmental issues.

—Standing

The legal concept of "standing" arises often in environmental cases. A person has "standing" when he/she has a right to have his/her case heard by a court. The concept of standing becomes important in environmental law situations when an environmental law is perceived as being either anti-industry or anti-environment and someone (either an individual person or an organization, for example) wishes to challenge the statute or regulation. The person seeking to sue must show that he/she is the intended beneficiary or a member of the group targeted by the offending statute or regulation and thus demonstrate that he/she has the legal right to bring such a challenge. Often, non-profit organizations seek to challenge a law on the basis that it is not protective enough of the environment and therefore the question becomes: "Who has the right to sue on behalf of the environment?" This interesting question is not merely academic: the answer can result in billions of dollars in compliance costs, legal fees, etc.

—Nuisance

Nuisance has existed as a legal concept for many years, beginning in the English common law. Essentially, the concept is defined in terms of a limitation on property rights: no one may use his/her property to impair the right of another person. This can mean something as obvious as keeping your trees from overhanging onto a neighbor's property to something as complex as preventing odors from an industrial plant from spilling over into the neighboring residential community. Most, if not all, municipalities have enacted ordinances which target nuisance-creating activities and conditions. A claim of nuisance has been one way that environmental issues have been brought within the legal system when there have not been set standards or criteria established via statute or regulation. The use of the nuisance concept in environmental law cases has, however, diminished as other legal concepts have become more accepted by the courts.

—Property Issues

Environmental regulation often affects the use of property, while state and the federal constitutions contain substantial protections of property owners' rights to use their property as they see fit. These competing concerns have led to conflicts and added new dimensions to old legal concepts. For example, one of the most prominent legal concepts addressing the restraints on the use of property is known as "eminent domain" or "taking," and reflects the right and power of the government to acquire privately owned property. Although this concept represents an action by the government to accomplish a particular purpose, the concept has also come to represent situations in which the government does not intend for a taking to occur, but yet a taking is found to have occurred. Most typically, a taking of private property occurs when new roads are being constructed or existing roads widened, although this power has also been used during wartime to appropriate properties deemed necessary by the government. These instances represent clear, intentional takings by the government to accomplish something related to the overall good of society. Yet, property owners have taken this idea and re-oriented it: private property owners have begun to question the extent of environmental regulations which limit the use of property by claiming that the government has effectively, albeit unintentionally, accomplished a "taking" of that property. A claim of "taking" by the government can be expected to be a part of any lawsuit challenging the applicability of a particular rule or regulation to a parcel of land. In such cases, the claimant's position is that the harshness and breadth of the rule or regulation which applies to his/her property has effectively prevented the property owner from having the use and enjoyment of his/her property to which he/she is entitled, and therefore, the government has robbed the property owner of his/her ownership rights and privileges. If the court concurs that a taking has occurred, the property owner is entitled to be compensated for the value of his/her property, or of that portion which has been affected.

There are other legal concepts and principles which originate in areas of the law other than environmental, but which have come to have a special significance in the environmental law context. In studying and understanding the science of the environmental concepts discussed in this book, it is useful to remember that our legal framework defines and confines those issues which affect society. Environmental issues remain among the most regulated and litigated, so an awareness of the legal framework will aid in gaining a more complete understanding of society's attempt to effectively address these concerns.

Consequently, Chapters 3 to 6 will also discuss how and why our society regulates the production and emission of selected substances and radiations that may be potentially harmful to humans. We will explore the issues involved in the production of energy and selected technologically useful substances and their disposal from the point of view of legislators, environmentalists, manufacturers, and homeowners. The role of the Environmental Protection Agency and the legal responsibility for the disposal of toxic wastes will also be described.

Science, Technology, and the Media

The media (television, feature-length films, radio, books, newspapers, and magazines) are an extensive source of information about science and technology for the public. Unfortunately, these same media are also the prime source of pseudo-science, namely false or grossly exaggerated depictions of science and technology. At the end of Chapters 3 - 6 are brief listings of "factual" television programs and other media sources of information. There are a number of feature-length films and television programs listed in Chapter 7 for which we will provide a brief plot description as well as the time at which the segment most closely associated with an environmental issue commences.

BIBLIOGRAPHY

Buck, Susan, J. *Understanding Environmental Administration and Law.* Island Press, Washington, DC, 1987.

Schwartz, Bernard. *The Law in America, A History*. McGraw-Hill Book Company, New York, 1974.

2
How to Research Environmental Issues: Science Research for Non-Scientists

College students who are not science majors need research sources that use language they can understand, that is, nonmechanical, non-scientific, non-jargon. These sources should lead the researcher to books, journal articles, and government reports that are intelligible to the non-scientist and that are readily available in a non-scientific academic environment.

This is not an impossible mission. The topics in this supplement—nuclear power plants and nuclear waste, indoor air pollution, stratospheric ozone depletion and the greenhouse effect, and the effects of electromagnetic waves on humans—have been discussed extensively over the past few years in every type of print and electronic medium, from tabloid newspapers to mass audience magazines to popular science periodicals to the most obscure physics and chemistry journals.

The research strategy in this chapter is designed to follow a path between the unacceptable tabloid and the incomprehensible scholarly scientific journal.

WHERE TO START
The hardest part of any research project is picking a topic. Four broad subject areas, all dealing with the science of specific environmental concerns, are designated for this supplement, eliminating the most difficult decision. Once one of these four broad areas is chosen, the next step in the research process is finding basic background information, written in non-scientific language.

BACKGROUND INFORMATION
No matter how long or short the paper, how narrow the focus, or how current the information must be, the best starting place is always an encyclopedia. Any major English-language encyclopedia, such as the *Academic*

American Encyclopedia, Encyclopedia Americana, Encyclopedia Britannica, Collier's Encyclopedia, or the *World Book Encyclopedia,* will provide the necessary background information. Encyclopedias answer a lot of basic questions about how long a topic has been studied, what population or community is affected, what other topics (broader or narrower) are of related interest, and the major books or journal articles to be consulted for further information. Use an encyclopedia to get a basic understanding of the subject, to help determine the aspect of a topic on which you will focus, and to figure out how far back in time to take a search.

Every academic library will have at least one major encyclopedia, in a print and/or electronic version. One of the most popular and commonly found electronic encyclopedias (published in CD-ROM format) is the *Information Finder*, from World Book. A search of the four major topics of this supplement yields the following articles: "Ozone Hole," "Ozone Layer," "Air Pollution" (sub-division: "Indoor Air Pollution"), "Nuclear Energy" (sub-divisions: "How Nuclear Energy Is Provided," "Hazards and Safeguards," "Wastes and Waste Disposal"), and, "Electromagnetic Waves." An advantage of an encyclopedia in electronic form is in the connections or "leaps" it permits, from broad to narrow topics, and from words or phrases found in the articles to other articles about those words.

DEPTH VS. TIMELINESS: BOOKS

Most professors set a required minimum number of sources for a research paper or report. For example, a five-page paper may require a minimum of one book, five magazine/journal articles, and perhaps one U.S. government report; a twenty-page paper may require five books, and 20 articles and reports. Finding one or more books for an in-depth review of a subject is easy: start with the titles that appeared at the end of the encyclopedia article. One book title from an encyclopedia article can lead to dozens of other sources.

Look up the title of the book in the library's card catalog or online catalog. If the catalog is electronic, look at the subject headings associated with that book, and next do a subject search on those subject headings. This turns up more books on the same topic. Look at the bibliography (source list) in the books found. These are the titles used by the writers of the books.

Books provide depth and quantity of information, and history of the issue. It is not logical or proper to write about the status of even the most current research topics without giving some background first. The present and future build on the past. Books provide the setting and background.

Books are also, by the very nature of their format, out of date before they are published. By the time a book has been researched, written, edited, published, and put on the shelves in a library, the information in it will be two to three years old. This information is too outdated to provide current material for a research paper on topics that are as timely as those covered by this supplement.

CURRENT INFORMATION: MAGAZINE/JOURNAL/PERIODICAL ARTICLES

Two- to three-year old information is just not current enough to describe the latest research, public opinion, and governmental action on environmental issues. Magazine/journal/periodical articles provide the most recent written research available. (The terms "magazine," "journal," and "periodical" are really interchangeable, although a magazine is usually considered less scholarly than a journal, which usually has footnotes and bibliography. If a requirement of the research paper is for scholarly journal articles, look for those that have footnotes and bibliographies.)

Wandering around in the shelves of current periodicals and hoping to find magazines with articles on the topic is not an efficient or effective research technique. What does work is choosing an index, in print or electronic form, that identifies, indexes, and (sometimes) abstracts the contents of magazines that cover the topic chosen.

There are thousands of commercially published indexes, in a wide variety of formats, covering every possible subject in the world. Listed below are some indexes, both print and electronic, that are guaranteed to provide an overwhelming number of sources on each of the four topics in this supplement.

Magazine Index Plus (electronic; InfoTrac software) and *Reader's Guide to Periodical Literature*

These two indexes, updated monthly, each cover several hundred popular magazines such as: *Environment, FDA Consumer, Popular Science, Science, Science News,* and *Scientific American.* These magazines are not considered "scholarly," and the articles rarely include footnotes and bibliographies, but the information in them will always be easily understood.

Three to four recent years of the *Magazine Index Plus* show several hundred articles on "Nuclear Energy," including "Environmental Aspects" of nuclear energy; on "Radioactive Wastes"; on "Indoor Air Pollution"; on "Ozone Layer Depletion"; on "Greenhouse Effect, Atmospheric"; and on "Electromagnetic Fields—Health Aspects." The sheer volume of information available on each of these topics in the popular literature is indicative of the strong public interest in environmental issues. The *Magazine Index Plus* includes brief abstracts of the articles it indexes.

The *Reader's Guide to Periodical Literature* is available in both print and electronic formats. This is one of the most familiar and popular periodical indexes, and it is found (in one format or another) in almost every school, public, and academic library. It covers almost the same journal titles as does *Magazine Index Plus*, uses similar subject headings, and will yield approximately the same number of articles.

Print and electronic versions of indexes have different advantages. Print indexes offer many cross references: if a subject word or phrase is not used by the index, there will likely be a referral to another subject or phrase that is used. An electronic index offers the possibility of searching several years of articles at once. Searching several years of the *Magazine Index Plus*, up to the current month's release, takes one search. *Reader's Guide* in print may cover the same time period with four annual volumes, plus several monthly paperback issues. Electronic indexes make it possible to search a longer time period easily; print indexes make it easier to search for only the most recent articles.

The overwhelming advantage of any electronic index is the normal inclusion of abstracts of the articles themselves. This makes it simple to decide whether or not the article is appropriate for the research project. It also allows for searches on key words or combinations of key words, a technique not possible in print indexes.

Another type of magazine or periodical that should be searched for information on these topics is newspapers. Newspaper articles are especially valuable for insight into current public opinion, mass media coverage and slant on an issue, and local, state, and federal legislative action taken or pending.

The National Newspaper Index (electronic; InfoTrac software)

This index is available in both an electronic version and a microfilm version. It covers five major U.S. national newspapers: *The Christian Science Monitor, The Los Angeles Times, The New York Times, The Wall Street Journal,* and *The Washington Post.* Since it is produced by the same company that makes the *Magazine Index Plus*, it uses the same subject headings.

Recent years of the *National Newspaper Index* show several hundred articles on "Nuclear Energy," on "Radioactive Wastes," on "Ozone Layer Depletion," on "Greenhouse Effect, Atmospheric," and on "Electromagnetic Fields—Health Aspects."

The three indexes described above will lead to hundreds of articles on any of the topics in this supplement. These articles will all be popular, easy to understand, and non-technical. To get a slightly more scholarly and scientific group of articles, try:

The General Science Index

The *General Science Index* (available in both print and electronic formats) is published monthly by Wilson, the same publisher that produces the *Reader's Guide to Periodical Literature*. This makes it a particularly familiar and comfortable index to use to track down popular science articles. The index covers about 130 English-language journals in all scientific fields, including atmospheric science, environment and conservation, medicine and health, and physics.

Among the journal titles covered (that are relevant to the topics of this Supplement) are: *Bulletin of the Atomic Scientists*, *The Conservationist*, *Environmental Science & Technology*, *Journal of Environmental Health*, Journal of the Atmospheric Sciences, *Natural History, Physics Today*, and *World Watch*. In one recent year alone *The General Science Index* listed more than 200 articles on "Radioactive Waste Disposal," on "Air Pollution, Indoor," on the "Greenhouse Effect," and on "Electromagnetic Fields—Biological Effect" and "Electromagnetic Waves—Biological Effect."

Because the journals covered by *The General Science Index* are slightly more academic than those in the *Reader's Guide*, the articles will usually fulfill a requirement for scholarly research sources. Look for the abbreviation *bibl* in the index listing of any specific article. This indicates that footnotes and/or bibliography are included in the article. These are considered scholarly articles.

All of the journals, magazines, and newspapers covered by the *Magazine Index Plus, Reader's Guide, National Newspaper Index*, and *The General Science Index* can usually be found in most large public and academic libraries.

All-in-One Index

It usually takes two to three separate research processes to collect enough books, magazine articles, and government reports to support a lengthy research paper. There is one index that covers all of these types of materials and is a good starting place for research. It covers such a wide variety of publications, from popular to scholarly to scientific, that it may be the only index needed to find information on these current topics.

PAIS International (electronic) and *PAIS International in Print* (monthly)

Previously called *Public Affairs Information Service Bulletin*, *PAIS International* and *PAIS International in Print* index and abstract new books, journal and magazine articles, popular U.S. and foreign government reports and documents, state government documents, and the publications of many important intergovernmental organizations (such as the United Nations, the World Bank, and the European Communities). Publications in six languages (English, French, German, Italian, Portuguese, and Spanish) are included, but all abstracts of foreign language publications are in English.

Many currently available indexes cover both books and journals in one subject field; a few also include government reports. In one year, *PAIS International* may index 1600 journals, 7000-8000 new books, and thousands of government documents.

PAIS International (in electronic format), in three to four recent years, lists 300-400 titles on "Atomic Power Plants," on "Nuclear Reactors," on "Radioactive Waste Disposal," on "Indoor Air Pollution," on "Atmospheric Ozone," on the "Greenhouse Effect, Atmospheric," and on "Electromagnetic Fields."

Many departments within the United States government are actively researching and writing about all of these issues. Government funds are allocated to deal with the problems, and laws are passed to control them. Including information from United States government documents in research papers on these topics enriches the reports. *PAIS International* can provide the easiest access to U.S. government publications.

The first four periodical and newspaper indexes described above are limited primarily to publications from and about the United States. This is not the only country concerned about and researching nuclear energy and wastes, indoor air pollution, ozone layer depletion and the greenhouse effect, and the effects of electromagnetic waves on humans. Consider adding an extra dimension to a discussion of these subjects by including the international perspective from the publications indexed in *PAIS International*.

Alternate Routes

To meet the demands for more scientific or scholarly content in a research paper, consider the following alternative indexes:

—*Environment Index* and *Environment Abstracts* and *Enviroline*

Updated monthly, these indexes are issued in print and electronic versions. They index and abstract a wide variety of publications and non-print materials. Included are new books, journals (mainly scholarly and scientific, some popular), conference proceedings, newspaper articles, radio and television programs, and material from the *Federal Register* (rules and regulations from United States Government agencies).

—*Pollution Abstracts*

Both print and electronic versions of this index cover the technical literature on pollution of the environment. Like *Environment Index*, it is narrow in subject focus, but covers the same wide range of materials, from books to journal articles to government reports and documents.

YESTERDAY'S HEADLINES AND LAST WEEK'S MAGAZINES

The problem of getting timely information on current topics is not completely resolved by using these magazine and journal indexes. Even if an index is updated and published monthly, the articles the monthly issue covers may lag six to seven weeks behind the current date. It takes a long time to acquire and index and abstract a title, and longer still to publish the index. Until only a few years ago, there was no effective way to find yesterday's headlines or last week's magazine articles. Serendipity and good luck were substitutes for methodical research.

Within the past few years the time between publication of an article and availability of a genuine index to it has been reduced to approximately 48 hours. In academic institutions with networked electronic indexes *and* access to a mainframe computer, one or more of these "instant indexes" may be available. Look for:

> *ArticlesFirst* from OCLC's *First Search*;
> *CARL Uncover*;
> *Eureka* from the Research Libraries Group.

The indexing service receives an issue of a journal, enters the table of contents into its mainframe database within 48 hours, and, instantaneously, the information can be searched worldwide by those connected and subscribed. The services of these vendors include document delivery—for a fee—within days (mail) or hours (fax). Answering the need for up-to-date research, these new indexes are an extraordinary development.

THE "INFORMATION SUPERHIGHWAY": THE INTERNET

The Internet, sometimes called "the Information Superhighway," is not an entity; it is a network of networks, mainframe computers speaking to mainframe computers around the world, allowing individual account holders to "surf" in search of academic research resources, professional colleagues and friends with similar interests, information and fun. It has become a mammoth self-perpetuating, self-governing, non-profit source of everything from e-mail groups, to weather forecasts, to statistical databases, to full-text government reports, to the catalogs of thousands of libraries around the world. It is increasingly being used for commercial purposes as well, but we will concentrate here on information sources that users can access at no charge.

Getting access to the Internet and its resources is different for each institution, each building, each piece of equipment. It is wise to take every introductory seminar offered about connecting to and using the Internet. For more information on the Internet, its background, use, and resources, see:

> Krol, Ed. *The Whole Internet: User's Guide & Catalog.* O'Reilly & Associates, Sebastopol, CA, 1997.

There are entire books written about the Internet in virtually every discipline. For environmental online resources, see:

> Feldman, Andrew J. *The Sierra Club Green Guide: Everybody's Desk Reference to Environmental Information.* Sierra Club Books, San Francisco, CA, 1997.

One current Internet resource for environmental information is the *EnviroLink Network*. To access it on the Internet, the online address is:

> **EnviroLink Network**   http://www.envirolink.org

For access to other environmental resource material, try some of the following resources:

> **U.S. Environmental Protection Agency**   http://www.epa.gov
>
> **United Nations Environment Programs**   http://www.unep.org
>
> *Amazing Environmental*
> *Organization Web Directory*   http://www.webdirectory.com

Please note: addresses of online resources are subject to change at any moment, without notice.

LEGISLATIVE UPDATE

Yesterday's crisis leads to today's governmental intervention and to tomorrow's tax burden. Government reaction to, and intervention in, an environmental problem can have an immediate impact on the daily life of the ordinary American. Jobs, salaries, lifestyles, health, homes, are all affected by environmental problems and by our government's handling of them.

Keeping up to date with the government's actions and tracking down legislative information is a full-time job in many libraries. There is one particularly good magazine, available in most academic libraries, that makes it easy to find out what issues (including environmental) are being discussed in Congress.

Congressional Quarterly Weekly Report

This weekly magazine, from a commercial (non-governmental) publisher, is one of the best known journals which follows the activities of the U.S. Congress. It provides narrative information on pending legislation, a "Status of Major Legislation" table, voting records of the members of the House of Representatives and of the Senate, and the text of remarks by political leaders.

Since U.S. Government publications are being issued more frequently in electronic format, and because they carry no copyright, the full text of an extraordinary number of government publications are available on the Internet. Try the Library of Congress online information system, *Thomas*, for more information from the United States Government.

Online address: **http://thomas.loc.gov**

THE FRUSTRATION FACTOR

Identifying what exists—what has been published, what Internet resources are available—is a fairly simple process. Finding or accessing them is trickier. Books are borrowed, magazines sent to binderies, and mainframe computers do crash. No more than 20% of sources identified (books, magazine articles, government reports, Internet links) will be accessible on the first search. (This is a corollary to Murphy's Law!) The frustration factor may be less at some institutions, but do not count on it.

The indexes and resources described in this chapter will lead to an overwhelming amount of information on any of the topics, **but make a bibliography (a list of resources) of five times as many sources as are needed for the report**. Identifying 20 sources and finding four of them is an outstanding success rate.

3
Nuclear Power Plants, Nuclear Waste Disposal, and Other Related Issues

Nuclear power plants are sources of electrical power that do not produce carbon dioxide and other greenhouse gases which contribute to global warming—the greenhouse effect—described in Chapter 5. However they pose other types of environmental dangers which will be discussed throughout this chapter.

As of November 1996, 110 nuclear power plants were operating in the United States. They generate approximately 20% of the electric power used annually in this country. Yet, no new plants have been ordered since 1976. Primarily, this is due to the issues of nuclear waste disposal and the safety of the plants. In addition, the operating licenses of 17 of these plants will expire by 2012. These licenses are not expected to be renewed. Thus, while nuclear power has the potential to replace fossil fuels as the main source of electric power, important environmental issues will first have to be adequately addressed.

The Basic Physics of Nuclear Power

The source of nuclear energy is the disintegration of a nucleus of uranium or plutonium. When a nucleus of the isotope uranium 235 (U-235) captures a neutron it is transformed into U-236, which is unstable and thus decays in a variety of ways. One of these decay modes is the division of the nucleus into two smaller nuclei with the emission of 2 or 3 neutrons and a good deal of energy. This is called nuclear fission. The energy released is due to the fact that the masses of the daughter nuclei and the emitted neutrons are less than the mass of U-236. The mass lost has been converted into pure energy using Einstein's $E=mc^2$ formula where E is the energy equivalent of a mass m and c is the speed of light in a vacuum. The amount of energy released is typically about 210 million electron volts, many millions of times greater than a chemical reaction involving an atom. (An electron volt = 1.6×10^{-19} joules.)

Uranium ore consists of only about .71% U-235, while the remaining ore is U-238 (99.28%) and U-234 (.01%). In order to use uranium in a conventional nuclear reactor the ore must be refined to increase the concentration of U-235 to about 3%. The nonfissionable U-238 can capture neutrons which reduce the number that can cause the fissioning of U-235. However, the slower the neutrons the greater the probability that U-235 will capture a neutron and the smaller the probability that a U-238 nucleus will capture it. Thus in a conventional reactor the neutrons undergo elastic collisions with a material (called a moderator) which slows them down without capturing too many of them. Most current reactors in the United States use water as a moderator.

Conventional Nuclear Reactors

The uranium fuel used in reactors is first processed into pellets, each a fraction of an inch long and a fraction of an inch in diameter. They are then imbedded in 12-foot-long fuel rods, which are then grouped into fuel assemblies, normally consisting of 200 such rods. A typical nuclear reactor has about 250 such fuel assemblies. Above each fuel assembly is a control rod, made of a metal alloy that absorbs neutrons. The plant operators can move the control rods up or down into the fuel assembly. If the control rod is pushed into the assembly, it absorbs many of the neutrons emitted in the disintegrations of U-235, and the reaction slows down. If the rod is pulled out of the assembly, it no longer absorbs the neutrons and more U-235 nuclei are struck by them and the process increases. Since the amount of heat produced by the reactor is proportional to the number of U-235 nuclei that are disintegrating, the heat production can be controlled by moving the control rods into or out of the fuel assemblies.

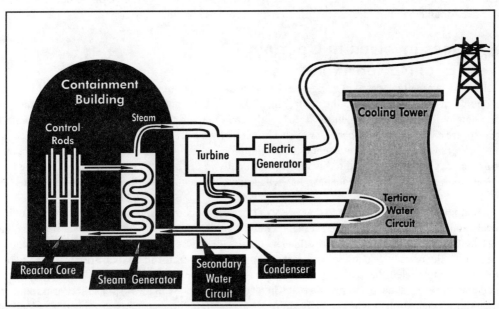

Figure 3-1: **A schematic of a conventional nuclear power plant.**

Figure 3-1 is a schematic of a conventional nuclear reactor. The fission of U-235 in the reactor core heats the water in the primary water circuit. This circuit is a closed system that circulates the hot water (at temperatures in excess of 500° Fahrenheit) at high pressures. Because the superheated water is at high pressure it cannot expand to become steam. This very hot water circulates from the reactor core to the steam generator where it converts the water in the secondary water circuit into steam. The pressurized steam then turns the turbine, which in turn spins an electric generator to produce electricity. The steam then runs to a condenser where it is cooled into a liquid again. The cool water is provided to the condenser by a tertiary water circuit. After the water in the tertiary water circuit is heated in the condenser it is recycled to the cooling tower, where it is cooled before being recycled back to the condenser. All United States commercial nuclear plants are light-water reactors, meaning that ordinary water is used to cool the fuel core. About two-thirds of these light-water reactors use pressurized water, as depicted in Figure 3-1. The remaining third of the reactors are boiling-water reactors in which the coolant water itself is allowed to boil, and the resultant radioactive steam directly drives the turbines.

Breeder Nuclear Reactors

These convert U-238, which constitutes 99.28% of uranium ore, into plutonium, Pu-239, a human-made isotope that is fissionable, just like U-235. Breeder reactors can thus use either U-235 or Pu-239 as their fuel. Because they can use U-238, they can generate much larger amounts of energy from a given amount of uranium ore. In essence, they can use as fuel part of the nuclear wastes from conventional reactors. The first breeder reactor experiments were conducted in the United States: however, breeder reactors have been developed mainly in Europe and Russia.

There are particular safety problems associated with breeder reactors. They must use liquid sodium instead of water as a coolant because the liquid sodium does not slow down the neutrons, unlike water. The faster moving neutrons are then more likely to interact with the U-238, transforming it into plutonium, which then splits apart providing the heat to turn water into steam in the secondary water circuit. However, liquid sodium is highly reactive (it reacts violently when exposed to air or water).

Furthermore, the plutonium which is produced in a breeder reactor is bomb-grade quality. It can easily be separated chemically from the uranium. It is feared that terrorists might obtain some of the nuclear products from a breeder reactor and then use these materials to build an atomic bomb.

Safety Issues Involved in Operating Conventional Nuclear Reactors

The reactor core is usually encased in a huge steel structure called the reactor vessel. This vessel is designed to prevent the release of radiation into the environment. In turn the reactor vessel and steam generator are built inside a containment building, which has steel-reinforced concrete walls 3 to 5 feet thick that were built to withstand severe earthquakes, hurricanes, or even a direct hit by a large jet airliner. The control room in a reactor has many devices which allow the personnel to both monitor and control what is happening in the reactor core. Some of these devices are designed to operate automatically in the event of a malfunction in some part of the reactor. But, as we shall see, accidents have happened in reactors, and some have been quite serious.

There are a number of issues relevant to the safety of operating a conventional nuclear reactor. The first set of issues deals with the design of the reactor. It is imperative that the reactor be surrounded by a strong containment building: the Chernobyl disaster, described later in this chapter, was due in part to the lack of a containment building around the reactor. Secondly, the personnel must be well trained and their training must be ongoing: human errors compounded the accidents at both Three Mile Island and Chernobyl. Finally, the responsible government agencies must have appropriate plans in place for dealing with a worst-case scenario requiring mass evacuation of the population around a reactor. In order to illustrate these issues, the nuclear accidents at Three Mile Island and Chernobyl will be described in some detail.

The Three Mile Island Accident

The most serious nuclear reactor accident in the United States occurred on March 28, 1979, at the Three Mile Island reactor located near Harrisburg, Pennsylvania. President Carter appointed a Commission to investigate the accident. In its report the Commission stated: "To prevent nuclear accidents as serious as Three Mile Island, fundamental changes will be necessary in the organization, procedures and practices—and above all—in the attitudes of the Nuclear Regulatory Commission and, to the extent that the institutions we investigated are typical, of the nuclear industry."

The Commission concluded that the fundamental problems related to this accident were "people-related problems and not equipment problems." Specifically, the incident was triggered when a pilot-operated relief valve at the top of the pressurizer failed to close when the pressure dropped. This created an opening in the primary cooling system. The indicator light in the plant's control room showed only that the signal had been sent to close the valve rather than the fact that the valve actually remained open. The operators relied on the indicator light and believed that the valve was closed rather than that it was open, causing a loss-of-coolant for two hours. The TMI emergency procedure for such a stuck open valve did not state that unless that valve was closed, a loss-of-coolant would exist.

The high pressure injection system came on automatically, as it should have: this was a major design safety system. However, the operators were concerned that the plant was "going solid," i.e., filling with water. Consequently, they cut back the high pressure injection from 1,000 gallons per minute to less than 100 gallons per minute. For extended periods on March 28, 1979, much of the reactor core was uncovered by water, resulting in severe damage to the core. If the high pressure injection had not been reduced by the operators, core damage would have been prevented despite the malfunctioning of the relief valve. If the operators or their supervisors had kept the emergency cooling systems on throughout the early stages of the accident, the Three Mile Island incident would have been relatively insignificant.

The Presidential Commission found that "the training of TMI personnel was greatly deficient. While training may have been adequate for the operation of the plant under normal circumstances, insufficient attention was paid to possible serious accidents." In addition, "specific operating procedures, which were applicable to this accident, are at least very confusing and could be read in such a way as to lead the operators to take the incorrect actions they did. Third, the lessons from previous accidents did not result in new, clear instructions being passed on to the operators."

Furthermore, the operators' problems were exacerbated by the fact that during the first few minutes of the accident more than 100 alarms went off in the control room. There was no system for suppressing the unimportant signals so that the operators could concentrate on the significant alarms. In short, insufficient attention had been paid to the interaction between human beings and machines under the confusing circumstances of an accident. The Commission concluded that "the most serious 'mindset' is the preoccupation of everyone with the safety of equipment, resulting in the down-playing of the human element in nuclear power generation. We are tempted to say that while an enormous effort was expended to assure that safety-related equipment worked as well as possible, and that there was backup equipment in depth, what the Nuclear Regulatory Commission and the industry have failed to recognize sufficiently is that the human beings who manage and operate the plants constitute an important safety system."

Just how serious was the accident? The reactor suffered a near-meltdown. The reactor will never be free of radioactivity and will eventually be entombed in concrete. But first, a cleanup was undertaken. The total cost of the accident is estimated to be well over three billion dollars. A second reactor at the plant was undamaged during the accident and was placed in operation in 1985.

During the TMI accident some radiation was released to the atmosphere, but the vast majority of radioactive material that had escaped from the core was prevented from reaching the outside world by the containment building. The amount of radiation received by any one individual outside the plant was low. However, even low levels of radiation may result in the later development of cancer or of birth defects among children who were exposed while in the womb. One calculation estimated that .7 cancers would result in the population affected by the radiation from TMI. But there will be no way of knowing if someone actually develops cancer among the 2,000,000 individuals (living within 50 miles of TMI) who may have had a minuscule exposure to additional radiation. Eventually, some 325,000 of these 2,000,000 are expected to die of cancer for reasons having nothing to do with the TMI accident. This number is only an estimate and the actual number could be 1,000 higher or lower. Hence, there is no statistical way of determining whether or not one additional person dies from cancer caused by radiation released by TMI.

On the positive side of the TMI accident, new safety regulations were established, including evacuation plans for the areas surrounding nuclear power plants. Finally, the complacency that had been common throughout the nuclear industry before TMI was also reduced. The nuclear industry formed its own watchdog agency, the Institute for Nuclear Power Operations (INPO). One independent group estimated that INPO has uncovered more safety problems at individual reactors than the NRC or the utilities themselves. The training of staff has also improved.

Safety Improvements in Reactor Designs

The nuclear power industry and governmental research laboratories have designed new types of reactors which they believe will be much safer to operate than the present ones. Some of these reactors will not have a meltdown even in the worst scenarios. Because the fuel cannot melt, radiation cannot be released into the atmosphere. Some designs propose using helium, an inert gas, rather than water to cool the system and turn the turbines. Helium is much less corrosive than steam, and is thus less likely to cause a leak in the pipes. The goal of these reactors is to be smaller, safer, and less expensive to build and to maintain. But gas-cooled plants also have problems. The nuclear power industries of both France and Great Britain initially were based on versions of the gas-cooled-reactor technology. The high capital costs and low reliability of the early gas-cooled plants caused both countries to switch to light-water reactors of the type used in the United States.

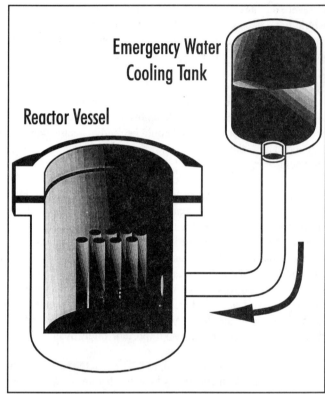

Figure 3-2: **A schematic of a proposed passive emergency water cooling system.**

Some of the new designs also reduce the number of mechanical systems needed for their operation. For example, some are designed to cool the reactor core "passively," using gravity instead of pumps. As Figure 3-2 demonstrates, in an emergency, cooling water would flow from the emergency cooling water tank into the reactor under the force of gravity, since the emergency tank would be positioned higher than the reactor. This would be safer than the present system in which emergency cooling systems rely on pumps with backup diesel generators.

Nuclear Plant Efficiency

As a result of the TMI accident, the U.S. government ordered changes in the way that reactors operated, resulting in the refitting of existing reactors and the hiring of hundreds of additional employees at most nuclear power plants. Operating and maintenance costs tripled during the 1980s. As a result, in 1991 the average nuclear power plant spent an estimated 5.54 cents to generate each kilowatt-hour of energy, about 85% more than the cost of a kilowatt-hour produced by a coal-fired plant. The percentage of time that the reactors are in operation also affects the cost of producing electricity, and this percentage is somewhat lower in the United States than in most European countries or in Japan, as Table 3-1 indicates. Some studies concluded that electricity produced from the most recently constructed nuclear power plants could cost up to twenty times as much as that produced from the first nuclear power plants. In addition, during the 1980s, the cost of construction of nuclear power plants increased to an amount three to five times that which had originally been projected.

Table 3-1: Percentage of Time Reactors are Working (1980s data)

Country	Number of Reactors	Percentage of Time
Finland	4	91.60
Hungary	4	86.60
Switzerland	5	84.50
Sweden	12	77.50
West Germany	21	73.00
Japan	28	71.20
United States	109	61.50
France	53	60.80
Great Britain	29	53.10

Because of these economic considerations, and the public's opposition to having a reactor built in their vicinity (nicknamed NIMBY—Not In My Back Yard), no new reactors have been ordered in the United States since the mid 1970s. Since the Chernobyl disaster in the former Soviet Union, European interest in constructing additional reactors has also waned.

One nuclear power plant, at Shoreham, Long Island, was never allowed to start operating despite construction costs of $5.5 billion dollars. The plant, completed in 1983, was prevented from opening on safety grounds by the governor of New York. The impasse lasted until 1989 when the plant was sold to the State of New York for $1. In return, the State pledged to allow the utility which had constructed the plant to raise customers' electricity rates for the next 10 years. It has since been partially dismantled: it is the nation's first nuclear power plant to be abandoned even before it was opened.

The Chernobyl Accident
The world's worst nuclear disaster occurred at the Chernobyl plant in the former Soviet Union on April 26, 1986, when one or two explosions ripped apart a nuclear reactor and released large amounts of radioactive materials into the atmosphere. Apparently some tests were being conducted at the Unit Four reactor at Chernobyl, in the Ukraine, about ninety miles from the large city of Kiev. Local fire fighters bravely battled the fire that had broken out after the explosion, and prevented it from spreading to other reactors at the power plant complex. Many of these fire fighters later died from overexposure to radiation. The fire was eventually contained by using helicopters to drop some 5,000 tons of sand and boron on top of the reactor to plug the leak.

Once the immediate danger had passed, the radioactivity at the plant had to be cleaned up as much as possible. The high levels of radiation made this also very hazardous. Robot-controlled vehicles malfunctioned, presumably due to the high levels of radiation to which the robots were exposed. Eventually, the Number 4 reactor was encased in 300,000 tons of concrete. However, these measures did not prevent the release of an enormous amount of radioactive debris into the atmosphere. As might be expected, the immediate vicinity was heavily contaminated. The areas immediately adjacent to Chernobyl, namely the Ukraine and Byelorussia (now Belarus), face long-term problems. About 20 percent of the farmland and 15% of the forests of Belarus cannot be used for more than a century because they are contaminated with radioactive materials with relatively long half-lives. The **half-life** of a radioactive material is the time that it takes for half of the radioactive nuclei to decay. At the end of one half-life there are only half as many undecayed radioactive nuclei and hence the decay rate is reduced by two. This means that the radiation per unit time received by an individual in the vicinity would also decrease by a factor of two.

Inhabitants in many parts of the Ukraine and Belarus cannot drink the water or eat locally produced fruits, vegetables, milk, and meat. No one knows how many people will ultimately die because of exposure to large amounts of radioactive fallout from Chernobyl. The health of about 350,000 persons in the Ukraine alone is constantly being monitored. Some estimate that 8,000 people in the Ukraine have already died as a result of the accident.

The Soviet Union was not the only country affected by the accident, however. Radioactive clouds from Chernobyl spread fallout over parts of Europe such as Sweden, Norway, Finland, Great Britain, Germany, Switzerland, and Italy. Parts of Poland and Hungary were also heavily contaminated. The unpredictability of where the radiation came to earth made it difficult to plan evacuation and other measures.

What caused this catastrophe? There are at least three contributing factors. The reactor itself, a Soviet model RBK-1000, is extremely unstable at low power levels. Secondly, the personnel were inadequately trained. Thirdly, the lack of a containment building spelled disaster at Chernobyl.

The Status of the Nuclear Industry in the Former Soviet Union

While the training of nuclear plant personnel was improved after the accident at Chernobyl and there were some safety improvements added to Soviet reactors, the present situation is still disquieting. With the disintegration of the Soviet Union, and the economic upheaval in Russia and the other countries that have nuclear reactors (29 in Russia, 14 in the Ukraine, 2 in Lithuania, 1 in Kazakhstan, and 2 in Armenia), it has become very difficult to obtain the parts needed for maintenance of these reactors. This is becoming critical since some of the plants are now 15 years old. Plant managers don't have the money to make safety improvements in their reactors, which by Western standards were not well-designed in the first place. In some plants, the life span of parts are being extended to a point that may be potentially dangerous. Workers are often not paid on time and their salaries are very low by Western standards. For example, in the Ukraine, plant operators are paid the equivalent of about $18 per month, which is low even in the Ukraine. Thus, trained workers are leaving the nuclear power industry and morale among those remaining is reportedly low.

Many of the nuclear power plants are owed substantial sums of money for the electric power they have provided. In Russia, most of the electricity is sold to a government monopoly which then sells it to individual customers. As of April 1994, only 30% of what was owed the power plants had been paid. They have been kept operating because their closing would be economically disastrous to the countries in which they are located. For example, in October 1993, Ukrainian officials rescinded an earlier decision to shut down the two working reactors at the Chernobyl power plant. These officials were undeterred by the reports that the sarcophagus entombing Reactor Number 4, which had been encased in 300,000 tons of concrete, is being eroded by the radiation inside it faster than expected. In Lithuania, 90% of electric power is generated by its two reactors. Perhaps the most dangerous of all is the decision announced by Armenian officials to reopen the nuclear power plant that is located in the same region in which an earthquake killed 25,000 people in 1988 and left 500,000 homeless. Both of the reactors to be reopened lack a containment building. The United States is providing financial assistance to Armenia to develop hydroelectric plants and to drill for oil. But these alternative energy sources will take many years to develop commercially, whereas Armenia has a pressing need for electrical power today. Thus, the need to alleviate dire economic conditions is taking precedence over environmental and public health concerns.

There are also concerns that the Russians may not be able to adequately safeguard the nuclear materials that are stored at numerous nuclear power plants and military facilities. In 1996 the United States spent about $400,000,000 to improve Russian nuclear security. This reportedly includes paying part of the salaries of 11,500 research scientists and staff in the former Soviet Union. These expenditures are being made to keep nuclear materials and nuclear weapons scientists from going to "third countries," where they could be employed to build weapons.

Decommissioning of Nuclear Reactors

Nuclear power plants can operate for only about 25 or 30 years. At that time the reactor vessel and other crucial sections become brittle or corroded. Then an important difference between nuclear reactor power plants and other power plants becomes clear. One cannot just demolish or abandon a nuclear reactor. Over its functioning lifetime, it has become contaminated with radioactivity. In addition, highly radioactive spent fuel rods, which were probably maintained in water-filled storage ponds while the plant was operating, must be disposed of safely.

There are three modes of retiring a nuclear power plant. The first mode is called storage. In this mode the plant is guarded by the utility company for perhaps 50 to 100 years, during which time some of the radioactive materials decay, making it less radioactive. This is not a permanent solution, but simply a way to make it easier to handle the plant under modes 2 or 3. However, the accidental release of radiation from the plant during this period is a constant threat.

The second mode is to entomb the plant in concrete. The problem with this as a permanent solution is that the tomb would need to remain intact for thousands of years. It is likely that accidental leaks would occur, as will soon be the danger from the Chernobyl reactor that was encased in 300,000 tons of concrete. Also, how can we guarantee that future governments would maintain the tombs?

The third mode of retiring a nuclear power plant is called decommissioning or dismantling the plant either immediately after it closes or after a period of storage. Some portions of the plant will be too radioactive for workers to handle, even if they wear protective clothing. Those sections will have to be torn down by advanced robot machines. These robots, however, will have to be built so as to overcome the problem the Russians confronted when they tried to use robotic machines in the cleanup of Chernobyl—namely, that the robots' circuits malfunctioned due to the high levels of radiation. After the plant is demolished, it is moved in small sections to a permanent storage site.

Several small nuclear power plants have already been decommissioned: Shippingport, America's first commercial nuclear power plant, was dismantled in 1989 and moved by barge more than 8,000 miles from its location in Pennsylvania to a military dumpsite in Washington State. However, the Shippingport reactor vessel was small enough to be shipped intact. For larger reactors, new robotic dismantling machines must be developed and a permanent storage site for radioactive materials must be established.

Decommissioning a nuclear power plant runs a number of risks. First, workers may be exposed to high levels of radiation and some radiation may escape into the environment during its dismantling. There are also dangers inherent in transporting the plant long distances to a permanent disposal site. This will be perhaps the major problem facing the nuclear power industry worldwide since there will be many retirement-age nuclear plants within the next two decades.

Disposal of Radioactive Wastes

Radioactive wastes are divided into two categories: low-level and high-level. Low-level radioactive wastes are solids, liquids, or gases that emit small amounts of ionizing radiation. They are, of course, produced by nuclear power plants, but they also originate from research programs and hospitals due to medical applications of nuclear physics. At present, they are stored in steel drums in three government-operated landfills.

High-level radioactive wastes are radioactive solids, liquids, or gases that initially give off large amounts of ionizing radiation. They are produced by nuclear power plants and nuclear weapons facilities and they are the most dangerous wastes produced by humans. They include spent fuel rods and assemblies as well as coolant liquids used in reactors. Some high-level wastes are created when neutrons are absorbed by the uranium fuel rods and radioactive isotopes are formed. Some of the isotopes remain radioactive for very long time periods. Plutonium-239 has a half-life of 24,400 years, neptunium-237 has a half-life of 2,130,000 years, and plutonium-240 has half-life of 6,600 years.

Clearly, permanent waste depositories must keep these materials from reaching the biosphere for hundreds of thousands of years. But, there is no entirely safe and permanent way of disposing of these radioactive wastes. Such storage sites must have geological stability and no ground water flowing nearby. At present, the best storage sites are believed to be those buried in deep salt deposits. Salt is an effective barrier to radiation and some salt deposits are believed to have been undisturbed for up to 200 million years, indicating that they are located in a geologically stable part of the Earth.

Today, there are no permanent storage facilities for radioactive wastes anywhere in the world. The Department of Energy therefore must build permanent waste depository sites to last 10,000 years. To understand the difficulty involved, one should consider that the oldest large-scale structure of human origin is the Great Pyramid in Egypt, which is about 4,600 years old. While 90% of the Great Pyramid still remains, 90% of a nuclear waste storage facility might not be sufficient to ensure that radioactive materials do not escape its confines. No government is likely to last 10,000 years, and even languages may change sufficiently that warning signs placed at the site may be unreadable. Furthermore, there is no way to insure that a cataclysmic event, such as a collision with a large meteor, will not rip apart whatever structure is used to contain these radioactive wastes.

There are about 100 sites in the United States which are designated as "temporary" storage sites for nuclear wastes. In 1982, the Nuclear Waste Act made the federal government responsible for selecting permanent waste storage sites and for having the first site operational by 2010.

The government has tentatively selected Yucca Mountain in Nevada as a permanent storage site for high-level nuclear wastes from commercially operated power plants. Similarly, it has tentatively selected Carlsbad, New Mexico, as the permanent waste storage site for nuclear wastes from the production of nuclear weapons. However, there are possible problems with both sites. Yucca Mountain is near a young volcano and active fault lines. An earthquake could both release radiation into the atmosphere and raise the water table, contaminating the groundwater. At the Carlsbad site, oozing brine could cause the steel drums, which are to be stored in rooms carved out of rock salt, to corrode and thus contaminate groundwater with radioactivity.

Even after the best sites have been selected and developed, there will still be hazards involved in transporting high-level radioactive wastes from the nuclear power and nuclear weapons plants to the permanent site. Many communities will object to these dangerous wastes passing near or through them. This is another aspect of the NIMBY (Not In My Back Yard) syndrome associated with determining a location for a nuclear power plant or nuclear waste site.

One method to minimize the need for permanent nuclear waste storage facilities is to consume a greater fraction of the nuclear fuel in the reactor by using breeder reactors which convert the abundant U-238 into plutonium and then use the plutonium as fuel. However, plutonium is used in nuclear weapons and the widespread construction of breeder reactors would substantially increase the difficulties in accounting for all nuclear fuel to prevent terrorists from building nuclear bombs.

Despite the difficulties in constructing appropriate permanent storage sites, they are desperately needed. Large areas near some nuclear processing facilities in both the former Soviet Union and the United States have been contaminated by planned or accidental discharges. About 3 million Curies (a Curie is 3.7×10^{10} nuclear disintegrations per second) of radioactive sources have been released in the United States while 1.7 billion Curies have been released in the former Soviet Union. Human exposure to sources at a level of tens of Curies can be very harmful. Clearly, further radioactive contamination of the environment must be avoided.

Nuclear Weapons

There are two types of nuclear weapons: the fission, or atomic bomb and the fusion, or hydrogen bomb. The **fission** bomb works as follows: when U-235 or Pu-239 undergoes fission both energy and 2-3 neutrons are released. If there is a sufficient amount of the fissionable material, called the critical mass, more than one of the released neutrons will strike other U-235 or Pu-239 nuclei which, in turn, will release other neutrons which strike even more nuclei, etc. This is called a **chain reaction** and is illustrated below in Figure 3-3.

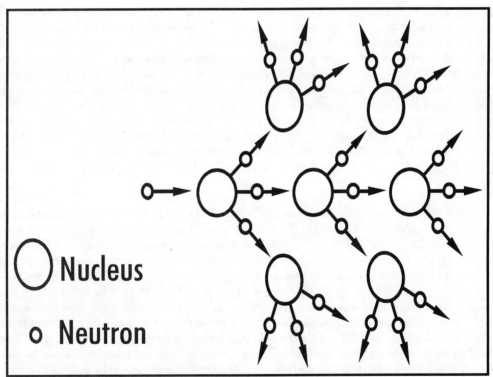

Figure 3-3: A chain reaction in which a disintegrating nucleus releases three neutrons, all of which strike other fissionable nuclei, causing their disintegration. The small circles represent neutrons and the large circles represent fissionable nuclei.

With each fission, energy is released. The entire process takes place in a fraction of a second and results in an uncontrolled nuclear reaction—an atomic bomb. The first atomic bomb was detonated in the deserts of New Mexico in 1945. Shortly afterwards, two atomic bombs were dropped on the Japanese cities of Hiroshima and Nagasaki, resulting in hundreds of thousands of casualties. Approximately 71,000 people were killed instantly at Hiroshima, and within 5 years, the number of deaths due to direct exposure to the bomb had reached an estimated 200,000. Almost 98% of Hiroshima's buildings were either destroyed or severely damaged. In Nagasaki, approximately 74,000 people were killed by the detonation, which destroyed 47% of the city. In the aftermath of this devastation, Japan surrendered, ending World War II. Each of the atomic bombs dropped on Japan produced an explosion equivalent to detonating about 20,000 tons of TNT.

In principle, an atomic bomb is easy to construct. High school students have designed atomic bombs! The weapon has two or more subcritical masses of U-235 or Pu-239 which are rapidly brought together by an explosive device. The one element in atomic bombs that is hard to obtain is the enriched U-235 or Pu-239. During World War II, the American effort to build a bomb was almost stymied by the difficulty of separating the non-fissionable U-238, which constitutes about 99.28% of uranium ore, from the U-235, which constitutes only .71%. Today, however, breeder reactors can provide an abundant supply of fissionable Pu-239. The critical mass necessary to construct a bomb out of Pu-239 would be a sphere the size of a baseball. Thus, an atomic bomb could be made quite small—the size of a suitcase—and would be relatively easy for terrorists to transport to a designated target. There is a practical limit to the explosive power of an atomic bomb of perhaps 100,000 tons of TNT since each of the nuclear masses must be subcritical and then all brought together simultaneously.

The **fusion** or hydrogen bomb works on a different principle. It fuses hydrogen nuclei to form helium with the release of energy. Actually isotopes of hydrogen—deuterium and tritium—are fused to make helium. Deuterium is present in water and is relatively easy to separate from normal hydrogen. Tritium is a human-made hydrogen isotope. It can be formed during the fusion reaction by bombarding another element—lithium—with neutrons. The isotope of lithium that is required, Li-6, is found in seawater. However, these isotopes of hydrogen will not fuse to form helium at ordinary temperatures because they all contain positive electric charges (protons) and positive charges repel one another. The attractive nuclear force only exists over very short distances and the hydrogen nuclei would be prevented by the electrical repulsive force from ever coming that close together unless they are moving very fast. The average speed of nuclei is equivalent to their temperature. The speed, and thus temperature, needed to place these hydrogen nuclei in close proximity is about 100,000,000 degrees Centigrade. One way of achieving this temperature is to detonate an atomic bomb. Thus, each hydrogen bomb has an atomic bomb at its core.

There is no limit to the destructive power of a hydrogen bomb. The largest hydrogen bomb tested to date had the explosive power of about 68,000,000 tons of TNT! America's B-52 bombers each could carry two bombs with an explosive power of 20,000,000 tons of TNT apiece. The destruction that could be caused by hydrogen bombs is difficult to imagine. Scientists have examined the probable results of an all-out nuclear war between the United States and the then Soviet Union. In addition to those killed by the blast and heat from the explosions, the entire world's population would be affected by the massive amounts of dust, debris, and smoke that would be produced both by the detonations themselves, as well as by the widespread fires they would ignite. This dust, debris, and smoke might screen out so much of the sunlight that the light available at noon would be equivalent to only the light of a full moon. Crops would fail and there would be world-wide famine. Numerous plants and animals would die in great numbers and some would even become extinct. Nuclear winter is the name given to this worst-case scenario. Other scientists do not think the environmental scenario would be quite as bad. There are uncertainties as to the optical properties of smoke particles and how high they would rise into the air and whether rainfall could remove most of these smoke particles from the air. However, it is certain that radioactive fallout would be widespread. Only insects would likely be the survivors since they are more resistant to the damaging effects of radiation than are many of their predators, such as birds.

The likelihood of all-out nuclear war with hydrogen bombs has been greatly reduced with the ending of the Cold War and the dissolution of the former Soviet Union. Hydrogen bombs are quite complex to build. At present only the United States, France, Great Britain, China, and Russia have hydrogen bombs (although who has ultimate control of hydrogen bomb-tipped missiles in some of the other former Soviet Republics, such as the Ukraine, is not entirely clear). Thus, civilization does not now appear to be in imminent danger of self-destruction in a global nuclear war fought with hydrogen bombs.

Unfortunately, there is an abundance of the materials necessary to produce atomic bombs. While an atomic bomb typically has a destructive power that is only 1/1000 that of a hydrogen bomb, it could still produce widespread death and destruction. The terrorists who attacked the World Trade Center in New York City used a bomb that had the explosive power of about one ton of TNT. Imagine if they had instead used a suitcase-sized atomic bomb with the explosive power of perhaps 5,000 tons of TNT. Both of the 100 story-high World Trade Center towers would have been destroyed, killing all 100,000 persons inside, and killing or injuring hundreds of thousands of others in densely populated Manhattan. America's leaders then would have a number of alternative responses, all unsatisfactory. There would have been nothing left at ground zero to examine for clues as to the bombers. Whom would we strike back against? What changes in our ability to move freely in and out of our country might also be sacrificed in an attempt to prevent another city from being devastated? The only "good" solution to the problem is to prevent it from happening. To do that, nuclear bombs must be kept out of the hands of terrorists. This will probably be civilization's major security challenge over the next couple of decades.

Since it is so easy to build a nuclear bomb once one has the nuclear fuel, the international community has concentrated its efforts on monitoring the production and use of nuclear fuel. Many countries have signed an international ban on the proliferation of nuclear weapons, but there are suspicions that some of the signatories are proceeding to acquire a nuclear arsenal nonetheless. The United Nations has the responsibility to inspect reactors in countries that are signatories to the non-proliferation treaty to insure that the fuel and waste materials from their commercial reactors are not diverted into the production of atomic bombs. There have been major international confrontations with Iraq, and most recently, with North Korea, over the operations of their nuclear energy programs.

With the adoption of a "Comprehensive Test Ban" by the United States, the nation must deal with certain problems in maintaining our nuclear weapons. Central to this issue are the following assumptions about our future nuclear weapons:

1. The United States intends to maintain a credible nuclear deterrent.
2. The United States supports world-wide nuclear non-proliferation efforts.
3. The United States will not develop any new nuclear weapons.

Since our policy is not to develop new nuclear weapons, our concerns are for the safety, security and reliability of our present stockpile of nuclear weapons. In particular, do we need to continue nuclear testing at around a yield of one-half kiloton of TNT in order to acquire critical information about the effects of aging on weapons in the nuclear arsenal? A study conducted for the Department of Energy during the summer of 1995 concluded that such testing was not crucial. However, the study also concluded that should the United States encounter problems in an existing stockpile design that lead to an unacceptable loss of confidence in either the safety or reliability of a weapon type, it is possible that tests with yields in excess of 10 kilotons of TNT might be needed. A "supreme national interest" withdrawal clause (standard in arms control agreements) would then have to be exercised to respond to the crisis.

The United States is currently dismantling 1,500 nuclear warheads per year and is manufacturing no new ones. After currently agreed to reductions of nuclear weapons (with Russia) are attained, the United States will still have about 3,500 nuclear warheads and associated launchers in its active strategic arsenal.

The Effects of Radiation On Humans

The term radiation includes X-rays, alpha particles (helium nuclei), beta rays (electrons), gamma rays (shorter wavelength electromagnetic radiation than X-rays), neutrons and protons. Any of these particles can ionize the atoms or molecules of materials they pass through: hence they are referred to as ionizing radiation. The damage to living organisms from radiation is due primarily to the ionization produced in cells. Ions can interfere with the normal operation of the cell which may thus die. The death of one cell may not be very serious; however, higher levels of radiation may kill so many cells that the organism dies. In addition, the radiation may damage the DNA of a cell so that it may survive but be defective in the way that it reproduces itself. The cell may go on dividing rapidly, producing more defective cells: the organism has then developed cancer due to exposure to the radiation. It is clear that there is some increased risk to any level of additional radiation. The challenge to science is to quantify the risk and to establish radiation dosage levels at which the risk is acceptable. A central principle in discussing the effects of radiation on humans is the assumption that it is better to receive a given amount of radiation in small doses over a long period of time than all at once. This is because humans have evolved over the ages exposed to low levels of radiation: hence, the human body is normally expected to be able to deal with low levels of radiation without developing cancer.

The unit of radiation is the **rad (radiation absorbed dose)**, which is the amount of radiation which deposits .01 joules/kilogram in any absorbing material. The rad is a unit of dosage. Another unit is used to measure the human biological damage caused by any radiation. This is because equal doses of different types of radiation do different biological damage. The **relative biological effectiveness (RBE)** of a given type of radiation is defined to be the ratio of the number of rads of X or gamma radiation that will produce the same biological damage as one rad of the given radiation. The RBE is one for X and gamma rays, one for beta rays, one for fast protons, 10 for fast neutrons and 20 for alpha particles. The most commonly used unit of radiation dosage is the rem (rad equivalent man). The effective dose in **rem** = the dose in rad x RBE. By this definition one rem of any type of radiation does the same biological damage.

The effects of exposure to radiation are immediately obvious: high-dose levels cause such symptoms as reddening of the skin, a drop in the white blood-cell count, nausea, vomiting, loss of hair, internal bleeding, etc. Changes in the white blood-cell count are detectable after a human receives a single 10 rem dose of radiation. A single 400 rem dose is fatal in about 50% of the cases. A single dose of 1000 rems is nearly always fatal.

The cells which grow the fastest are the most susceptible to damage from radiation. Thus, the bone marrow cells, which produce blood cells, are particularly susceptible. Other sensitive tissues are the ovaries, the testes, and the lenses of the eye. During childhood, when tissues are actively growing, exposure to radiation will often have a more severe impact than during adulthood. Health problems in children exposed to high levels of radiation include growth, development, and organ dysfunction, hormonal deficiencies, and cognitive impairments.

Radiation is also used to *treat* cancer because cancer cells generally grow faster than normal cells and thus are more susceptible to destruction by radiation. A cancer patient's radiation therapy may consist of crossed beams of radiation that pass through his/her body such that all of the beams intersect at the site of the cancer. Alternatively, the patient's body may be rotated while it is irradiated in order to reduce the damage to the non-cancerous cells in the path of the radiation beam. Radiation can also be used to diagnose illness. In addition to X-rays of the body, radioactive tracers can be used to plot blood flow, for example.

Typical and Maximum Permissible Radiation Doses

A 1993 Report of the United Nations Scientific Committee on the Effects of Atomic Radiation presented the following data:

Table 3-2: Annual Effective Doses for Adults from Natural Sources

Cosmic Rays	.039 rem
Terrestrial Gamma Rays	.046 rem
Inside Human Body	.023 rem
Radon and its decay products	.130 rem
TOTAL:	**.238 rem**

In addition, medical X-rays in the United States, on average, contribute perhaps another .09 rems to the annual radiation received by Americans. Other sources of radiation, including nuclear power plants and nuclear bomb tests, contribute less than .01 rems per year. The total annual radiation dose for adults in the United States is thus in the vicinity of .3 rems. It should be noted that these are just averages. The dosage from radon will vary with its location and the level of ventilation in a given home. The amount of radiation from cosmic rays increases with the elevation of one's home since the higher up the home is located, the less atmosphere remains above the home to shield against cosmic rays. By the same reasoning, an airline pilot receives additional radiation from cosmic rays of about 0.0005 rems per hour while flying. This may result in an additional 0.5 rems of radiation per year. Similarly, one might receive much larger dosages from medical X-rays than the average amount quoted above.

The maximum rate recommended for the general public by the Environmental Protection Agency is 0.5 rems per year. For radiation workers, the maximum permissible dosage is presently 5 rems per year divided into four quarters in each of which a worker is limited to no more than 1.25 rems.

Recent studies on nuclear workers exposed to low levels of radiation confirmed that reasonably accurate extrapolation of mortality rates for exposure to low levels of radiation can be made from groups exposed to high dose rates, such as survivors of Nagasaki and Hiroshima. Thus the current radiation protection recommendations for environmental and occupational exposures are probably accurate, although they are based in part on extrapolations from high dose rates.

Radiation Experiments Conducted On Humans

In 1986, the House Subcommittee on Energy Conservation and Power issued a report, "American Nuclear Guinea Pigs: Three Decades of Radiation Experiments on U.S. Citizens," which documented 31 government-sponsored radiation experiments on 695 people from the 1940s to the early 1970s. In 1994, Energy Secretary Hazel O'Leary ordered the declassification of millions of pages of documents, some of which describe yet other radiation experiments. The numbers of persons involved in those experiments and the level of radiation to which they were exposed were described in subsequent news stories. Congress has already started compensating victims of these experiments. In 1990, it passed the Radiation Exposure Compensation Act (PL 101-426), which provides $50,000 for residents who lived downwind from certain nuclear tests and later developed certain kinds of cancer. In 1993, the Act was amended to provide $75,000 for any soldier who developed cancer after participating in above ground nuclear war tests. A trust fund of $150,000,000 has been established for residents of the Marshall Islands who became ill or lost their homes as a result of nuclear bomb tests in the 1940s and 1950s.

Operation Desert Storm and Depleted Uranium Weapons

Depleted uranium (DU) is a by-product of the uranium enrichment process used to supply fuel for nuclear reactors. It has a lower content of fissionable material than natural uranium. It is also extremely dense and thus can be used in the armor of tanks and other combat vehicles as well as in the production of shells to penetrate armored targets. Depleted uranium is both radioactive and a chemically toxic material (much like lead).

During the 1991 Persian Gulf War, named Desert Storm, 29 U.S. combat vehicles were contaminated after being hit by DU rounds from Abrams tanks (some of which have DU in portions of their armor) or having stored DU rounds explode from other accidents during combat. At least several dozen U.S. soldiers were exposed to DU by inhalation, ingestion or shrapnel as a result of these events. Tests on some of these soldiers indicate that their exposure was likely within the safe limits set by the Nuclear Regulatory Commission for radiation exposure and toxicity. However, an investigation by the General Accounting Office indicated that soldiers had worked in and around DU-contaminated combat systems without being aware of the characteristics of DU ammunition, the potential risks from DU contamination, or the precautions necessary to minimize DU exposure. The Army has promised to improve its training of military personnel with respect to DU equipment.

Irradiated Food

Recently, the use of radiation to kill microorganisms in food has been undertaken commercially on a widespread scale. Proponents of this method of sterilizing food note that medical supplies have been sterilized in this manner for some time and they claim this process is perfectly safe. It also has economic benefits such as increasing the "shelf-life" of foods that have been irradiated. Opponents question whether the effects of irradiating the foods are non-hazardous. About the only thing that is certain is that the public will be hearing about this debate for some time to come.

SOCIETY'S REACTIONS TO THESE ENVIRONMENTAL ISSUES

Like other energy-generating processes, nuclear power represents both a benefit and a risk to society. Justifying the presence of any risk in society requires a weighing of the risks and the benefits, a process which is a function of government. If the government decides that the benefits outweigh the risks in general, it then sets about to offset and limit those risks through legislation and regulation. Thus, while nuclear energy has many current uses in the United States, each use is accompanied by a myriad of requirements and licensing which represent "protection strategies" designed to preserve the health and safety of people and the environment.

Unlike other energy sources, nuclear power had its origin within the context of war and was created for a destructive purpose, rather than for the energy it could produce in peacetime. The circumstances in which nuclear energy was unleashed had a great deal to do with how the government regulated its use, and how society has viewed nuclear power over the years.

History of Regulation

The mechanism by which nuclear energy and radioactive materials are regulated consists primarily of the federal law known as the Atomic Energy Act of 1954,[1] as amended, and its associated regulations. This Act replaced the Atomic Energy Act of 1946[2] whose enactment was prompted by the creation of the atomic bomb in 1945 during World War II and the use of nuclear power by the government in war (that is, the production of nuclear weapons).

As is often the case with new products or technology developed during war-time, the origin and early years of nuclear energy were the sole province of the federal government. Certain sections of the 1946 Act, for example, gave the Atomic Energy Commission exclusive ownership of nuclear reactors,[3] fissionable materials,[4] and the exclusive authority to control the dissemination of technical information relating to nuclear weapons and reactors.[5] Once the peace-time uses of nuclear energy became apparent, there arose a need to regulate the generation and use of nuclear energy in a context which was broader than war weaponry. This represented the stimulus behind the passage of the 1954 Act.[6]

It wasn't until 1959, however, that Congress authorized the NRC to delegate certain aspects of nuclear regulation to the individual states.[7] In other words, no longer did the federal government have exclusive province over nuclear energy creation and use. Such delegation was not unusual as there has always been an emphasis on allowing individual states to regulate activities within their own borders, providing that the regulatory activities do not conflict with federal legislation or regulation.[8] Since 1959, states have been delegated significant authority to regulate such issues as by-product materials,[9] radioactive air pollution from nuclear power plants,[10] siting and land use requirements of nuclear plants,[11] and technical qualifications for the operation of nuclear plants.[12] States have also attempted to wrest control over other nuclear activities away from the federal government by challenging the federal government's failure to delegate this authority. This struggle between the states and the federal government has yet to be resolved but has resulted in numerous court cases.[13]

The 1946 Act had created the Atomic Energy Commission to promulgate regulations for the government's use of nuclear power and materials. The principal change in the 1954 legislation was the expansion of the role of the Commission to include the promotion of the civilian use of nuclear energy[14] and the regulation of all nuclear power activities. The Commission was subsequently abolished in 1974 when the government decided to split the Commission's functions between two entities when it passed the Energy Reorganization Act of 1974.[15] Functions relating to nuclear weapons and nuclear energy production were vested in the Energy Research and Development Administration (ERDA). The ERDA's functions were then vested in the Department of Energy when that agency was created in 1977.[16] Regulatory functions became vested in the Nuclear Regulatory Commission (NRC). Many of the current regulations governing the production of nuclear energy, including the possession, storage and use of radioactive materials and wastes and the construction and use of devices in the generation of nuclear energy, were issued by the NRC.[17]

The Handling, Transporting, and Disposing of Nuclear Materials

The use of nuclear power to create energy creates a myriad of other issues, all of which are regulated, as well. Other federal laws overlap with the Atomic Energy Act in the regulation of nuclear material in certain circumstances, such as situations when the public may be exposed to radiation (Consumer Product Safety legislation)[18] and those times when nuclear material must travel (Hazardous Materials Transportation Act).[19] The NRC shares regulatory authority over the transportation of radioactive materials pursuant to the Hazardous Materials Transportation Act and its associated regulations, which are promulgated by the Department of Transportation.[20] Every aspect of the transport of radioactive materials is regulated: labeling, manifesting, packaging, type and quantity transported, and method of transportation. These federal statutes place the transportation of nuclear material under the authority of the federal government and pre-empts all state and local requirements as to most aspects of transportation,[21] although some limited involvement by the states is permitted, such as permitting the individual states to designate specific highway routes for the transportation of the material.[22]

The final, but perhaps most important issue raised by nuclear energy in society, is how to satisfactorily address the disposal of radioactive waste and by-products. While state and federal regulations direct what cannot be done with such material, there is also substantial guidance on what *can* be done. Interestingly enough, the federal government specifically assigned the responsibility for the disposal of low-level nuclear waste generated by power plants to the individual states in the Low-Level Radioactive Waste Policy Act of 1980.[23] This Act "encourages" the management of such waste on a regional basis, among adjacent states. The states are directed to work together to reach agreement on waste disposal in their "compact" and choose appropriate sites for disposal. In 1985, the Act was amended to provide further "encouragement" in the form of deadlines, incentives and penalties.[24]

The disposal of so-called high-level nuclear waste is regulated by the Nuclear Waste Policy Act of 1982, as amended,[25] which keeps the majority of control at the federal level while authorizing states and Native American tribes to participate in the selection of an appropriate disposal site. This site, which is intended to be a permanent repository for the waste, will be constructed and operated by the Office of Civilian Radioactive Waste Management, a division of the Department of Energy,[26] and funded by the nuclear power industry through contributions to the Nuclear Waste Fund.[27]

RELEVANT EDUCATIONAL AUDIO-VISUAL MATERIALS

Note: In some cases, A/V materials can be previewed before purchasing or renting.

Acceptable Risks. (1987) 54-minute color documentary film examining the town of Canonsburg, PA, where tons of radioactive waste were buried in the town center, and residents are dying of cancer. Produced by Channel 4, England. Available from Filmmakers Library, Inc., 124 East 40th Street, New York, NY 10016, (212) 808-4980.

Atomic Energy: Science Fiction - Science Fact. (1988) 25-minute color videotape. Available from: Queue, Inc., 338 Commerce Drive, Fairfield, CT, 06430, (203) 335-0908 / (800) 232-2224 / FAX (203) 336-2481.

Ground Zero Plus Fifty: A Half Century of Nuclear Experimentation and Devastation. 52-minute color videotape. Available from: Films for the Humanities and Sciences, P.O. Box 2053, Princeton, N. J. 08453-2053, (609) 275-1400 / (800) 257-5126 / FAX (609) 275-3767.

Inside a Nuclear Plant. 58-minute color videotape. A look inside the largest nuclear reprocessing facility in the world, at Sellefield, England. Available from: Films for the Humanities and Sciences, P.O. Box 2053, Princeton, N.J. 08543-2053, (609) 275-1400 / (800) 257-5126 / FAX (609) 275-3767.

Nuclear Nightmare Next Door. (1990) 52-minute color documentary produced by CBS as part of the *48 HOURS* newsmagazine series. Explores the controversy surrounding the use and disposal of radioactive materials, the health effects on workers and nearby residents, and some ordinary citizens organizing against the dangers nuclear power and weapons plants are creating for their families and communities. Available from: Films for the Humanities and Sciences, P.O. Box 2053, Princeton, N.J. 08543-2053, (609) 275-1400 / (800) 257-5126 / FAX (609) 275-3767.

Nuclear Power Plant Safety: What's the Problem? 60-minute color videotape. Available from: Films for the Humanities and Sciences, P.O. Box 2053, Princeton, N. J. 08453-2053, (609) 275-1400 / (800) 257-5126 / FAX (609) 275-3767.

NOVA: Back to Chernobyl. (1989) 60-minute color documentary film from the public television series, *NOVA*. Available from: Queue, Inc., 338 Commerce Drive, Fairfield, CT, 06430, (203) 335-0908 / (800) 232-2224 / FAX (203) 336-2481.

Radioactive Waste Disposal, the 10,000 Year Test. 50-minute color videotape. Available from: Films for the Humanities and Sciences, P.O. Box 2053, Princeton, N.J. 08543-2053, (609) 275-1400 / (800) 257-5126 / FAX (609) 275-3767.

Suicide Mission to Chernobyl. 60-minute videotape. Available from: Films for the Humanities and Sciences, P.O. Box 2053, Princeton, N.J. 08543-2053, (609) 275-1400 / (800) 257-5126 / FAX (609) 275-3767.

INTERNET RESOURCES

Access the following sites on the World Wide Web for further information (Note: URLs are subject to change.)

Radiation and Health Physics
http://www.sph.umich.edu/group/eih/UMSCHPS/Source.htm
This web site discusses nuclear radiation effects on humans.

Nuclear Regulatory Commission
http://www.nrc.gov
This web site belongs to the Nuclear Regulatory Commission. It discusses nuclear reactors, nuclear materials, radioactive waste, and rule making.

BIBLIOGRAPHY

Bradley, J. B., C. W. Frank, and Y. Mikerin. "Nuclear Contamination From Weapons Complexes in the Former Soviet Union and the United States," *Physics Today*, April 1996, page 40.

Candis, Elisabeth. "Effects of Low Dose Protracted Exposures to Ionizing Radiation: Nuclear Worker Studies," *Physics and Society*, January 6, 1997, page 6.

"Congress Ponders Compensation for Radiation Test Subjects," *CQ*, January 8, 1994.

"The Legacy of Three Mile Island," *The Philadelphia Inquirer*, March 27, 1994, page D1.

"Lethal Legacy: Soviet Nuclear Reactors," *The Philadelphia Inquirer*, June 13, 1994, page 1.

Marples, D. R. *Chernobyl and Nuclear Power in the USSR*, Macmillan Press, 1986.

"Nuclear Power is Losing its Steam," *Insight* magazine, January 15, 1990, page 46.

"Operation Desert Storm: Army Not Adequately Prepared to Deal With Depleted Uranium Contamination," *Report to the Chairman, Subcommittee on Regulation, Business Opportunities, and Energy, Committee on Small Business, House of Representatives*, GAO/NSIAD-93-90, 1993.

Pollack, C. "Decommissioning Nuclear Power Plants," from *State of the World*, ed. L. R. Brown, W.W. Norton and Co., New York, 1986.

The President's Commission on The Accident at Three Mile Island, Pergamon Press, New York, 1979.

Public Utilities Fortnightly, Nov. 15, 1996, page 10.

Raven, P. H., L. R. Berg, and G. B. Johnson. *Environment*. Saunders College Publishing, Philadelphia, 1995.

"Russia's Nuclear Stockpile: How Safe is it From Theft?," *The Philadelphia Inquirer*, April 18, 1996, page A1.

"Russia's Rising Nuclear Peril," *The Philadelphia Inquirer*, June 12, 1994, page A1.

Schwade, S. "Radioactive Muscle? The Glowing and Not-So-Glowing Reports on Food Irradiation," *Muscle & Fitness* 53: 8, August 1992.

Slovic, P., M. Layman, and J. H. Flynn. "Lessons From Yucca Mountain," *Environment* 33:3, April, 1991.

"Suffering in the Shadow of a Nuclear Plant," *The Philadelphia Inquirer*, June 14, 1994, Page 1.

Sullivan, J.D. "End of Nuclear Testing," *Physics and Society* 26, April 4, 1997, page 4.

United Nations Scientific Committee on the Effects of Atomic Radiation. *Sources and Effects of Ionizing Radiation*, United Nations, New York, 1993.

Worsnop, R. L. "Will Nuclear Power Get Another Chance?" *Congressional Quarterly's Editorial Research Reports*, February 22, 1991.

DISCUSSION QUESTIONS

1. What are the differences in construction and operation between a conventional nuclear reactor and a breeder nuclear reactor?

2. Why is it important to have a containment building around a reactor?

3. Why was the Three Mile Island accident far less detrimental to the environment than the Chernobyl accident?

4. Describe some suggested safety improvements in nuclear reactor designs.

5. Which country has the most "efficient" nuclear reactors?

6. What are the potential dangers in the nuclear industry in the former Soviet Union?

7. What are the risks involved in decommissioning a nuclear power plant?

8. Describe the NIMBY attitude towards nuclear power.

9. Draw a schematic of an atomic bomb.

10. What are the differences between a hydrogen bomb and an atomic bomb?

11. Why is radiation hazardous to human health?

12. Where would you expect to receive a higher level of radiation, at sea level or on the top of a high mountain?

13. How much occupational-related radiation might an airline pilot receive during a year?

14. Describe the present guidelines for radiation experiments conducted on humans.

15. Why was depleted uranium used in constructing some Abrams tanks?

16. Research the present regulations on the manufacture and sale of irradiated food.

NOTES

1. 42 USC Sec. 2011-2096, as amended.

2. Pub. L. No. 79-855, 60 Stat. 755.

3. 42 USC Sec. 2011.4(c)(1).

4. 42 USC Sec. 2011.5(a) (2).

5. 42 USC Sec. 2011.10.

6. According to the Senate Report accompanying the 1954 Act, the 1946 Act had considered that atomic energy was 95% a military issue and only 5% a peace-time issue.

7. Pub. L. No. 86-373, 73 Stat. 688 (1959), 42 USC Sec. 2021.

8. Article IV of the United States Constitution.

9. 42 USC Sec. 2021.

10. 42 USC Sec. 7422, part of the Clean Air Act amendments of 1977, Pub. L. No. 95-95, 91 Stat. 720.

11. 42 USC Sec. 2133-2134, Nuclear Regulatory Authorization Act for Fiscal Year 1980.

12. Senate Rep. No. 870, 86th Congress 1st session 2, reprinted in *1959 U.S. Code Congress and Administration News* 2872, 2879.

13. See, for example, *Pacific Gas and Electric Co. v. State Energy Resources Conservation and Development Commission*, 461 U.S. 190 (1983), *Silkwood v. Kerr-McGee Corp.*, 464 U.S. 238 (1984).

14. See S. Rep. No. 1699, 83rd Congress Second Session (1954), reprinted in *1954 U.S. Code Congress and Administrative News* 3456-59.

15. Pub. L. No. 93-438, October 11, 1974, 42 USC Sec. A5801-5891.

16. Pub. L. No. 95-91, 91 Stat. 577 (1977), 42 USC Sec. 5814, 42 USC Sec. 7151.

17. 10 CFR Parts 11, 20, 30, 33, 40, 73. The regulatory power was pursuant to 42 USC Sec. 2201(p).

18. 42 USC Sec. 2636n and 42 USC Sec. 10001-08.

19. 42 USC Sec. 1801-12.

20. See 42 USC Sec. 170, 2106, 49 USC Sec. 103, 104, 106, 1471, 1472, 180-13, 10 CFR Part 71, 49 CFR Part 170-89.

21. 49 USC Sec. 1804(a) (1988).

22. 49 USC Sec. 1804(b).

23. 42 USC Sec. 2021b - 2021j.

24. Pub. L. No. 99-240, 99 Stat. 1842 (1985).

25. 42 USC Sec. 10101 - 10270 (1988).

26. 42 USC Sec. 10224.

27. 42 USC Sec. 10222.

4
Indoor Air Pollution

Over the last 40 years, a great deal of governmental attention has been devoted to the analysis and correction of outdoor air pollution. The major urban and industrial pollutants have been linked to effects on human and animal health, accelerated corrosion or deterioration of materials, reduction of visibility, and climatic changes. Less attention has been paid to indoor air pollution, despite the fact that the manifestations are strictly human health degradation. Since the average person typically spends in excess of 90% of his time indoors, there is substantially more contact with indoor air pollutants.

Some of the most commonly encountered forms of indoor air pollution are illustrated on the following two pages in Figure 4-1. Included are smoke from cigarettes and stoves, radon gas, asbestos fibers in some older buildings, a variety of chlorinated hydrocarbon gases and vapors, plus living microorganisms and their by-products. In the United States, the three forms of indoor air pollution producing the highest death rates are radon, cigarette smoke, and asbestos. Radon tends to be a cold-climate problem where buildings are frequently sealed to prevent heat loss. Cigarette smoke may be encountered in any society, while asbestos is associated with the older industrialized countries. In the less developed countries, smoke from tobacco and stoves constitute the major indoor air pollutants.

Radon

Radon gas is a naturally occurring, radioactive element which is dispersed widely through rock and soil. Radon has several isotopic forms, the most common being the Rn^{222} form derived from the radioactive decay of uranium 238. Among the elements which make up the Earth, there are three radioactive materials (U^{238}, U^{235}, Th^{232}) which have half-lives on the order of billions of years and which are found in small quantities in most geological formations. Each of these atomic forms decays through a complex chain of radioactive emissions until the material becomes a non-radioactive or "stable" lead. In the decay process, a number of electrically charged alpha and beta particles are ejected from the nucleus of the decaying isotope along with gamma radiation.

Although the radioisotopes represented in the major decay series are spread through rock and soil, higher concentrations are associated with ore bodies, granite, igneous intrusions, shale, and phosphate rock. Radon-222 is the most commonly encountered isotope due to the relative abundance of its parent, Uranium-238. Radon-220 from Thorium-232 is occasionally observed.

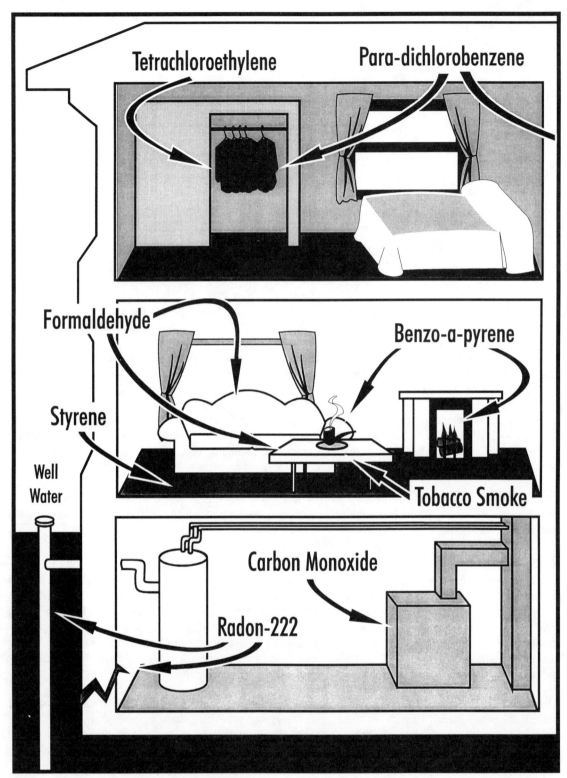

Figure 4-1a: Some major indoor air pollutants

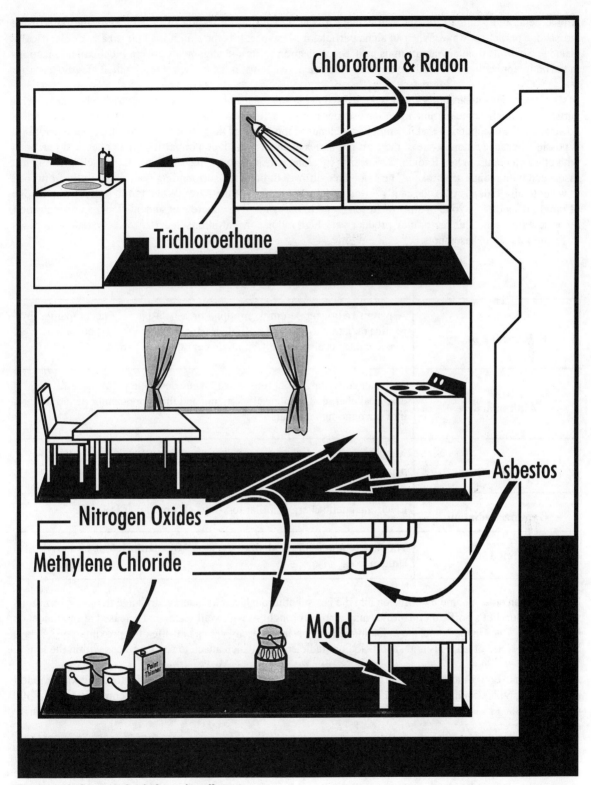

Figure 4-1b: **Some major indoor air pollutants**

Alpha, beta, and gamma radiation can damage tissue by causing ionization of molecules in the path of the radioactive particle or wave. When an alpha particle, for instance, hits the skin it may physically or electrically dislodge electrons from any atoms in its path. Since electron shifts and alignments govern biochemical reactions in the body, ionization will produce unwanted chemical reactions in the body, some of which may be harmful.

Most of the radioisotopes in rock and soil are fixed in place. Alpha and beta emissions are trapped in the mineral particles, although some gamma radiation escapes to form a very low-level field within which we live. Radon, however, is the exception in that it is mobile and tends to move out of the ground into the atmosphere. Mobility is possible because radon is one of the inert gas series elements which do not participate in chemical reactions with other elements. When Radium-226 decays by alpha emission to produce Radon-222, any chemical bonds associated with radium are broken. The radon then diffuses through the surrounding matrix generally from rock to water to air. Radon is thus constantly diffusing from the surface of the Earth. When radon moves from a zone of rapid diffusion to a zone of slow diffusion (sand to clay or soil to concrete basement floor) a concentration increase will occur. This factor may produce very high radon concentrations in certain well waters or in gases diffusing into some basements.

Table 4-1: A Radiation Lexicon

Radioactive Decay	A parent radioisotope emits an alpha or a beta particle accompanied by gamma radiation to produce a daughter which may or may not be radioactive. The daughter is always a different element than the parent.
Alpha Particle	A cluster consisting of 2 protons and 2 neutrons. Due to its size and electrical charge, it is intensely ionizing and therefore cannot penetrate far into surrounding materials.
Beta Particle	An electron ejected from the nucleus of a decaying atom. Typically can penetrate a few millimeters.
Gamma Radiation	Electromagnetic energy similar to x-rays except higher frequency.
Half-Life	The time required for one-half of the atoms of a radioisotope to decay. Unique to each isotope.

Radon gas can enter a home by diffusion into the basement through cracks, sumps, drainage trenches, or porous walls. Smaller, but significant quantities may also be brought in with well water and released at the faucet or shower head. Once in the home, the radon immediately starts to decay with a half-life of approximately 3.8 days to produce a series of lead, bismuth, and polonium radioisotopes which attach to solid surfaces within the home.

Although, on average, radon emissions from home wells is considered a small part of the total indoor air radon exposure, individual wells may contain as much as one thousand times the maximum concentration limit of 300 pCi/l established by the EPA for "public" water supplies. In such a case a shower can become an extraordinarily radioactive event.

The principal health impact of radon gas in a home results from lung damage due to alpha particles released by radioactive decay while the radon is in the lungs. Since radon is non-reactive, most of the radon entering the lungs is simply breathed back out without harm. Roughly one radon atom out of 100,000 breathed will decay while in the lungs. However, the alpha particle emitted is in intimate contact with lung tissue and the radon decay produces other isotopes which will emit future alpha and beta particles.

The EPA has estimated that 20,000 cases per year of lung cancer result from exposure to radon in the home. The total mortality is by far the highest of any known form of indoor air pollution. Some 85% of the lung cancer cases associated with high radon-level homes ocur in persons who also smoke. Thus the correlation between occurrence of the disease and radon concentration is complex. However, the EPA has estimated risks as shown in Table 4-2. Within the range of values shown, the EPA recommends that action be taken to reduce concentrations exceeding 4 picocuries/liter (pCi/l). Within the 4 to 20 range, action is recommended within several years. As levels approach 200, the continuation of exposure should not exceed several months.

Table 4-2: Indoor Air Risks Evaluation

Radon Risk Evaluation of Indoor Air

pCi / l	Estimated lung cancer deaths due to radon exposure (out of 1000)	Comparable exposure levels	Comparable risk
200	440 - 770	1000 times average outdoor level	more than 60 times non-smoker risk
100	270 - 630	100 times average indoor level	4 pack-a-day smoker
40	120 - 380		2000 chest x-rays per year
20	60 - 210	100 times average outdoor level	2 pack-a-day smoker
10	30 - 120	10 times average indoor level	1 pack-a-day smoker
4	13 - 50		5 times non-smoker risk
2	7 - 30	10 times average outdoor level	200 chest x-rays per year
1	3 - 13	average indoor level	non-smoker risk of dying from lung cancer
0.2	1 - 3	average outdoor level	20 chest x-rays per year

It is interesting in the political sense that no federal regulations have been adopted restricting allowable radon concentrations in homes. Extensive regulations and large expenditures have been mandated to protect outdoor air quality despite the smaller predicted human health impact. Presumably, the difference relates to a perceived resistance on the part of homeowners to what would amount to personal mandates.

Radon restrictions are increasingly being adopted on a local government level and real estate agents often recommend measuring radon and reducing concentrations to 4 pCi/l or less to avoid potential future liability associated with the sale of a potentially hazardous residence.

Smoking

Smoking is one of the more hazardous activities of the human race. World-wide, some 2 to 2.5 million people die annually of the smoke-induced diseases consisting of lung cancer, emphysema, bronchitis, and certain heart diseases. In the United States, tobacco is estimated to cause 325,000 deaths per year at a total increased health and insurance cost of between $38 and $95 billion per year, or $1.25 to $3.15 per pack of cigarettes sold. From the indoor air pollution perspective, non-smokers are exposed to the hazardous constituents of tobacco smoke which may build up to substantial concentrations in enclosed and poorly ventilated areas. In 1985, the EPA estimated a mortality rate due to passive smoke of 5,000 persons per year in the United States. The National Research Council, in 1986, also estimated that spouses of smokers face a 30% increased risk of lung cancer and that children of smokers face a 20% to 80% increase in risk of respiratory disease.

Although many people object to tobacco smoke because of its odor, the harmful ingredients are primarily fine particles which may deposit in airways or the lungs and organic tars which contain a variety of carcinogenic or otherwise toxic compounds. The upper respiratory tract has excellent filtration mechanisms for particles above 10 micrometers in diameter, but finer particles, which are common in smoke, can penetrate freely.

Increasingly stringent regulations at all governmental levels are restricting the areas in which smoking is permitted. Congress has considered legislation that would tax cigarettes at roughly $2 per pack in order to recover the increased health costs imposed on society by smokers. The tobacco industry, however, has been remarkably successful in minimizing and delaying restrictions on tobacco consumption. State lawsuits to recover health care costs for smokers may change this record.

In the United States, the percentage of people smoking dropped from approximately 40% in the mid-1960s to under 25% two decades later. Unfortunately, the numbers have stabilized somewhat and in many less developed countries, are actually increasing. A substantial part of the U.S. tobacco crop is exported.

Smokers who quit return to non-smoker risk levels in times ranging from one year for heart disease to less than 15 years for lung cancer. For non-smokers exposed to normal indoor air pollution levels, one would expect residual risks to subside very quickly after exposure was terminated.

Asbestos

Asbestos is the third major indoor air pollutant known to produce significant mortality rates. Asbestos is a small family of naturally occurring minerals which can be separated into small-diameter fibers. Asbestos is chemically stable, electrically resistant, and conducts heat poorly. These properties have led to historically widespread uses as woven insulation blankets, heat- and fire-resistant cemented sheets, reinforcement for roofing shingles, and temperature-resistant electrical insulators. Short fragments of asbestos fiber can be deposited in the lung where they may induce lung cancer or mesothelioma after lag times of 20 years or more. Permissible or safe levels of exposure have not been established since very low concentrations have been known to cause disease. The families of some asbestos workers have contracted mesothelioma from contact with fibers brought home on hair or clothing.

The EPA estimates that between 3,000 and 12,000 persons die per year in the U.S. due to asbestos-related forms of cancer. Most of these deaths are among former asbestos workers who were exposed to relatively large dosages of fibers. From the general indoor air pollution perspective, asbestos should be a decreasing problem as asbestos-containing materials and machinery are replaced. Congress passed the Asbestos Hazard and Emergency Response Act of 1986 which mandated the removal of asbestos-containing materials from schools.

Asbestos may still be encountered in certain older homes and buildings in the form of asbestos-filled vinyl floor tile, ceiling tile, old insulation on pipes or boilers, asbestos shingles for siding, or asphalt roofing shingles. It is important to note that the materials are only harmful if the fibers are inhaled. Siding shingles may liberate fibers if crushed or by deterioration, but it is very difficult to separate fibers from a vinyl floor tile. Probably the most hazardous form of asbestos is woven fiber mats sometimes encountered as insulation on old boilers and associated with steam piping.

The removal of asbestos-containing materials from a building is a complex task. Detailed regulations have been developed concerning isolation of work area, air filtration, and protective gear for workers. Materials removed are generally handled as hazardous wastes. Frequently, the most cost-effective solution is to seal the asbestos-containing materials under a protective coating.

Other Harmful Substances

Many potentially harmful chemical vapors may also be found in the home, generally in very low concentrations. These may be grouped into chlorinated organics, benzene ring-containing molecules, and other reactive chemicals such as formaldehyde and chloroform. Most are present in very low concentrations and are therefore unnoticed except occasionally by a faint odor. Where health effects occur, susceptible individuals are affected but most people are not. The effects are frequently encountered symptoms such as headache, dizziness, nausea, etc.

Chlorinated compounds encountered in buildings include:

Tetrachloroethylene - Commonly used by many (but not all) dry cleaners. Some traces remain after the clothes are dried and may volatilize slowly in the home. In high concentration, causes damage to nerves, liver, and kidneys.

1,1,1 Trichloroethane - A volatile and relatively inert chemical widely used in aerosol spray cans as the propellant. May cause dizziness and irregular breathing.

Methylene chloride - A very effective softener for paint residues and therefore widely used in paint removers. Evaporates very readily. Associated with nerve disorders and diabetes.

Commonly encountered benzene ring-containing organic compounds include:

Para-dichlorobenzene - Most commonly encountered in mothballs. Evaporates (sublimes) slowly at room temperature to maintain vapor concentrations. Potential carcinogen.

Benzo-a-pyrene - Present in wood and tobacco smoke. Potent carcinogen.

Styrene - Unreacted residues from styrene plastics used in carpets and various plastics. Causes kidney and liver damage.

Vinyl chloride - Unreacted residues in vinyl plastics. Implicated in liver cancer.

Chloroform has been detected in hot shower water, presumably from the reaction of chlorine added to drinking water with methane from microbial decomposition. Other chlorinated compounds are present, but they are not usually volatile. Chloroform is carcinogenic in high concentration.

Formaldehyde is used in certain plastics and adhesive resins, usually in combination with urea. Plywood and particleboard may contain formaldehyde glues, the unreacted portions of which can evaporate slowly over time. Foam insulation used in buildings is also often formaldehyde-based due to the low flammability and rigidity of the material. Formaldehyde is an eye, nose, and throat irritant and may also cause nausea or dizziness in some individuals.

Certain inorganic gases normally associated with outdoor air pollution may occasionally be found in buildings. Oxides of nitrogen (NO_x), including the toxic form nitrogen dioxide (NO_2), are produced from combustion or high temperature processes whenever temperatures exceed approximately 1500° F (817° C). Automobiles and industrial boilers and heaters produce most of the oxides of nitrogen present in outdoor smog. Indoors, NO_x is produced by unvented natural gas or kerosene heaters, by faulty furnaces, and by leakage from fireplaces. Nitrogen dioxide is a respiratory irritant, particularly in children.

Carbon monoxide (CO) is a product of incomplete combustion of any carbon-based fuels. Automobile exhaust is the principal outdoor source. Within a building, carbon monoxide may be produced by any inefficient or smoky fire. Unvented or improperly vented natural gas or kerosene heaters can malfunction and produce very high concentrations of CO in enclosed spaces. A number of people are killed each year in this manner. Normally, heaters, furnaces, and fireplaces will produce lower concentrations which, while not deadly, may cause headaches and drowsiness.

Ozone (O_3), an irritant gas usually associated with photochemical smog, is produced by certain office copy machines. In most cases, the ozone decays rapidly back to oxygen. However, significant concentrations may occur in a poorly ventilated copy room.

Microorganisms of several types may contribute to indoor air pollution. In a widely publicized event, a number of persons attending an American Legion convention in Philadelphia in the 1970s became seriously ill and several died. After intensive investigation, a bacteria subsequently named Legionella Pneumophila was found growing in the hotel's air conditioning cooler system. The water containing the bacteria was used to chill the hotel air and some bacterial aerosols were carried into the ventilation system. The disease has recurred occasionally at other locations, including a cruise ship in 1994.

A much more common problem results from mold growing on damp surfaces within a building. Mold produces trace quantities of volatile organic vapors which give the characteristic "musty" odor associated with dampness. Some people experience allergic reactions to the odor.

Sick Building Syndrome

The term "sick building syndrome" has been coined for those buildings which produce adverse effects on their occupants. Normally the term is applied to office buildings in which workers experience an abnormal incidence of headache, nausea, dizziness, eye irritation, sore throat, etc. In what may be the most famous case of all, the EPA completed a new headquarters building in 1988. Upon moving into the building, workers experienced a variety of health problems. Extensive investigation attributed the cause to organic vapors exuding from the new carpeting. Eventually the carpeting was replaced and the problem was solved.

Any type of building occupied by humans may be considered "sick" if it produces adverse health effects. In 1985, the EPA estimated that an individual was three times more likely to obtain cancer from the chemicals in a home than from breathing polluted outdoor air. Where the effect is dramatic, it is easy to diagnose a sick building; but where limited numbers of perhaps sensitive individuals are affected, even recognizing the problem may be difficult. Once a building is suspected of being sick, pinpointing the cause can be difficult and expensive. The EPA has estimated that 20% to 30% of U.S. buildings may be sick.

The problems of indoor air pollution and sick buildings has in some ways become worse during the last 25 years. In the wake of the Arab oil embargo of the 1970s, energy conservation became a universal goal. In ordinary homes, newer windows and doors have been designed for minimal leakage of air, which is the primary source of dilution for indoor pollutants. Building codes require reasonable ventilation in large buildings, but little or none in smaller commercial buildings and homes. Since the economics of energy is unlikely to change, the solution to indoor air pollution cannot rely on simple air dilution.

Indoor Air Pollution Controls

In Table 4-3, the generic approaches to pollution control are linked to the appropriate types of pollutants previously discussed. Ventilation, or passing controlled amounts of outdoor air through a building, is probably the most fundamental technique for pollution concentration reduction. On a home scale, however, mechanical ventilation is not economical. Instead, leakage is relied upon to supply fresh air. Larger buildings will have air cleaners, recirculation and make-up air supplies. In such cases, problems generally occur in poorly ventilated portions of the building or when the make-up air is reduced to cut energy costs. Ventilation of home basements using heat exchangers to conserve energy is often recommended to reduce very high radon levels. The cost, however, is substantial for small systems.

Table 4-3: Indoor Air Pollution Control

Pollutant	Ventilator	Exclusion or Removal	Product Selection	Maintenance	Notes
Radon	✔	✔			Exclusion most common
Cigarette Smoke	✔	✔			
Asbestos		✔	✔	✔	
Toxic Organic Chemicals	✔	✔	✔		
Inorganic Bases	✔			✔	
Microorganisms				✔	Dehumidify to prevent mold

Exclusion or removal entails preventing a pollutant from entering a building or removing its source. For radon, the most common control techniques are to block the gas diffusion through openings in the basement or foundation wall and/or to collect the radon by applying suction to the area beneath the floor slab or outside the foundation walls. "Vacuuming" the radon away from the walls is effective but requires continuous operation of an air blower. Cigarette smoke is increasingly being excluded by bans on indoor smoking but may also be removed by air cleaners. Small electrostatic precipitators to collect smoke have been common in bars for many years. Asbestos has been removed from most schools while the toxic organic pollutants are probably best controlled by eliminating their use in occupied spaces.

The selection and use of non-polluting products is a universal goal where technology permits. Manufacturers, with government encouragement, have made substantial progress in making products less harmful by substituting relatively safe ingredients. Polymer products such as rugs and vinyl plastic are treated to reduce unreacted residual chemicals that would vaporize after manufacture. Plywood and particle board have been altered to reduce the potential for formaldehyde production. Many products used in the home have also been reformulated to reduce toxicity. "Pollution prevention," or the elimination of pollution problems by not using harmful materials or products, is increasingly being emphasized by regulatory officials as the most effective cost-control technique.

Maintenance of materials and equipment is important in minimizing indoor air pollution. Asbestos-containing materials are only a problem if the material is allowed to deteriorate. Something as simple as a coat of paint or plastic resin can radically reduce an asbestos hazard. Heaters of all types also require maintenance to insure peak combustion efficiency and the prevention of leaks into interior spaces.

The control of microorganisms growing in building air conditioning systems is readily achieved by maintaining adequate concentrations of an appropriate algicide/bactericide in the water or by periodically adding chlorine. However, if those materials come in direct contact with air entering the building, it could be argued that the added chemicals may create a separate air pollution problem. On a residential scale, the common problem is mold resulting from water leaks or dampness. Repair of leaks and operation of a dehumidifier, particularly in a basement, will normally control the growth of mold.

SOCIETY'S REACTIONS TO THESE ENVIRONMENTAL ISSUES

As mentioned earlier, despite the increasing awareness of the health risks posed by indoor air pollution, regulation aimed at minimizing or eliminating such risks is virtually non-existent. The legislation that does exist is not cohesive or extensive. At the federal level, particularly, indoor air pollution programs are "fragmented" and insufficiently staffed and funded.[1] There is no legislative program which defines a consistent approach; in fact, no federal agency regulates indoor air pollution *as such*; certain specific indoor air pollutants, however, such as radon,[2] asbestos[3] and chlorofluorocarbons[4] have been regulated extensively. Cigarette smoking is widely regulated as well, although at the state level only.[5] Many individual states also regulate asbestos, radon and other indoor air pollutants, although no state act may be less restrictive than the federal acts regulating the same substance.

That is not to say that the federal government has not considered the passage of comprehensive legislation. In 1990, the Indoor Air Quality Act of 1990 was passed by the Senate, but not acted on by the House.[6] Likewise, the Indoor Air Quality Act of 1993 was introduced, but not enacted.[7] Considering that some commentators have termed indoor air pollution the "predominant issue of the 1990s,"[8] passage of such legislation may be merely a matter of time.

In general, the states have been no more active than the federal government; individual approaches can best be described as scattered, with activities variously including the enactment of new regulations to address specific pollutants, enactment of new regulations to address general indoor air concerns and the use of existing regulations to incorporate indoor air pollution issues. Much of the regulation has centered on informing the public of the dangers of indoor air pollution. Few states have passed comprehensive legislation addressing indoor air pollution issues in their entirety.[9]

Application of Existing Laws to the Indoor Air Pollution Issues

In the absence of specific government action in the form of new legislation, certain private groups have undertaken to develop guidelines for addressing indoor air pollution within certain contexts, such as in the construction industry.[10] Nevertheless, there has not been a widely accepted, interdisciplinary consensus standard embraced by the public. Aside from development of voluntary standards and short of enacting new legislation to specifically address these issues, there is support for looking to existing statutes to regulate indoor air pollution.

The most obvious and seemingly logical legislation to look to is the federal Clean Air Act,[11] which authorizes the federal Environmental Protection Agency (EPA) to regulate pollution which "enters the ambient air."[12] Interestingly enough, however, the EPA has limited the application of the definition of "ambient air" to "outdoor air."[13] To date, no challenge, judicial or otherwise, has been brought to test the agency's interpretation of the statute. This interpretation has left the most logical source of regulation unavailable to address indoor air pollution issues.

Another statute which could be interpreted as being applicable to indoor air pollution is the Toxic Substances Control Act (TSCA),[14] which was enacted to prevent "unreasonable risk of injury to health or the environment"[15] from "chemical substances." The definition of "chemical substances" however may be too narrow to support the inclusion of certain more complex pollutant compounds within the purview of the statute. In any event, EPA also administers TSCA, and the federal agency has not attempted to expand or interpret the statute to address indoor air pollutants.

The Occupational Safety and Health Act (OSHA)[16] provides another potential source of legislative guidance and standards. This statute requires that employees not be exposed to conditions in the workplace which represent a "significant risk" to worker safety or diminish worker health, functional capacity or life expectancy. Such conditions include pollutants in the air of the workplace and the OSHA regulations include limits on the concentrations of specific contaminants in the air, such as asbestos and certain volatile organic compounds. OSHA's authority has not been interpreted as extending, however, to all indoor air pollutants, possibly due to the "significant risk" requirement.

The Consumer Product Safety Act (CPSA),[17] which, like OSHA, is not administered by the EPA, authorizes the Consumer Product Safety Commission to regulate consumer products and directs the promulgation of safety standards applicable to the sale and use of consumer products. The definition of consumer products,[18] however, raises an interpretational difficulty similar to TSCA: the definition of "consumer products," while broad, does not include explicit reference to "building materials," a typical and likely source of pollutants. Even assuming that building materials are included in the statutory definition, this certainly would not constitute a comprehensive regulatory approach to indoor air pollution issues.[19]

The federal Comprehensive Environmental Response, Compensation and Liability Act (CERCLA, also known as "Superfund")[20] has also been suggested as a source of authority for regulating certain aspects of the indoor air pollution problem. CERCLA imposes liability for cleanup of hazardous substances and wastes, wherever found, on each and every person responsible for said substances or wastes or the property on which they come to be located. Liability under the statute is broadly defined and has been interpreted by the courts as extending to situations involving releases of hazardous substances into the air within a building.[21] In general, however, the courts seize on air issues to characterize a particular event as a "release" for the purposes of triggering the statute's broad liability scheme. There has been no judicial attempt to define CERCLA as regulating the indoor air environment.

Common Law and Other Approaches

The shortage of legislation which directly mandates controls on air pollution in buildings does not prohibit those who have been injured by pollutants from seeking relief. Although a regulatory framework provides the best foundation for ensuring that risks from indoor air pollution are minimized or eliminated, reliance on generally accepted legal principles (that is, the "common law") may accomplish the same end result in a litigation context. Common law principles which may be applicable include breach of contract, breach of warranties, both express (such as a written warranty) and implied (such as the implied duty to construct a safe building), strict liability, negligence, fraud or misrepresentation, nuisance and infliction of emotional distress.[22] In general, litigation brought by persons injured as a result of exposure to indoor air pollutants have been complex and troublesome, in part due to the difficulty in proving the absence of other causes of the alleged injury.

Even if successful, however, such cases rarely force a change in the responsible party which would prevent future exposures; instead, the outcome of successful litigation is usually in the form of a money payment to "compensate" for the injury to the claimant's health. In fact, a defendant in litigation may opt to pay a money settlement rather than have a case proceed to court where it could result in a court decision which establishes damaging precedent for the defendant's industry. Such tactics further prevent real change from being instituted in the way indoor air pollution is addressed and only emphasize the need for legislation.

Another source of relief for a person injured by indoor air pollution may be afforded by state workers' compensation law. The disability of an employee caused by exposure to pollutant(s) may result in compensation through the administrative claims process outlined by such regulatory framework.

Future Controls

It seems clear that society will ultimately require a cohesive, encompassing approach to indoor air pollution, likely through federal legislation. Only through such a regulatory framework will the issues have the potential to be adequately resolved. The current "piecemeal" approach to regulating indoor air pollution issues falls drastically short of reducing the risk posed by exposure to such pollutants.

RELEVANT EDUCATIONAL AUDIO-VISUAL MATERIALS

Note: In some cases, A/V materials can be previewed before purchasing or renting.

Air Pollution: Indoor. (1989) 26-minute videotape. Available from: Films for the Humanities and Science, P.O. Box 2053, Princeton, N.J. 08543-2053, (609) 275-1400 / (800) 257-5126 / FAX (609) 275-3767.

Breath Taken. (1989) 33-minute videotape exploring the effects of asbestos on workers. Available from: Fanlight Productions, 47 Halifax Street, Boston, MA 02130, (617) 524-0980 / (800) 937-4113 / FAX (617) 524-8838.

Radon. (1989) 26-minute videotape. Available from: Films for the Humanities and Sciences, P.O. Box 2053, Princeton, N.J. 08543-2053, (609) 275-1400 / (800) 257-5126 / FAX (609) 275-3767.

Radon Free. (1990) 32-minute videotape. Available from: Xenejenex, 300 Brickstone Square, Andover, MA, 01810, (508) 475-3000 / (800) 228-2495 / FAX (508) 475-0909.

INTERNET RESOURCES

Access the following sites on the World Wide Web for further information. (Note: URLs are subject to change.)

EPA: An Introduction to Indoor Air Quality (IAQ)
http://www.epa.gov/docs/iedweb00/index.html

This site provides introductory information on common indoor air pollutants, including:
- radon [Rn]
- environmental tobacco smoke [ETS] (secondhand smoke)
- biologicals (e.g., molds, bacteria, dust mites)
- carbon monoxide [CO]
- organic gases (VOCs) (e.g., solvents, cleansers and disinfectants, dry-cleaning fluids)
- formaldehyde (e.g., pressed wood products, furniture, fabrics)
- pesticides
- asbestos
- lead [Pb]
- nitrogen dioxide [NO2]
- respirable particles

The site also discusses sources of indoor air pollution; IAQ controls and measures in homes, offices/large buildings, and schools; provides information on "Indoor Air Pollution and Your Health"; and lists numerous resources for further information.

EPA Air Pollution Prevention and Control Division
http://www.epa.gov/docs/crb/index.html

This site is the homepage of the Air Pollution Prevention and Control Division (APPCD) of the National Risk Management Research Laboratory. It is EPA's major research organization for research, development, and evaluation of air pollution control technologies. The Lab has six branches, one of which is the Indoor Environment Management Branch (IEMB). Clicking onto this link will provide the user with a great deal of information about research projects conducted by IEMB in order to improve understanding of the relationships between indoor air quality (IAQ) and emission sources, heating, ventilating and air conditioning (HVAC) systems, and air cleaning devices. Research topics include: indoor air pollution sources; pollution prevention techniques; ventilation and air cleaning; cost analysis of IAQ control techniques; radon mitigation research, and more. In addition, the site provides information about a newsletter, upcoming symposia, and other resources in the field of indoor air pollution management.

Occupational Safety and Health Administration (OSHA)
http://www.osha.gov/

The mission of the Occupational Safety and Health Administration (OSHA) is to save lives, prevent injuries and protect the health of the more than 100 million American workers and six and a half million employers who are covered by the Occupational Safety and Health Act of 1970.

OSHA's site includes FAQs (frequently asked questions) on topics such as asbestos removal and indoor air pollution, as well as links to numerous other state and federal sites related to safety and health, including:

NIOSH (National Institute For Occupational Safety And Health)
http://www.cdc.gov/niosh/homepage.html

U.S. Department of Health and Human Services
http://www.os.dhhs.gov

World Health Organization (WHO)
http://www.who.ch

One of WHO's many resources is the *Health and Environment Library*. See the module on *Air Quality* (http://www.who.ch/programmes/peh/gelnet/hlm97air.htm). WHO has been concerned with air pollution and, in particular, its dangers to human health, for 30 years and it has produced a large number of studies on the various aspects of this issue. These studies cover not only the increasing urban air pollution caused by power generation, motor vehicle traffic, residential heating and industry, but also information on indoor air pollution both in developed and developing countries where the use of biomass fuel for domestic purposes puts particularly women and children at risk. The module is composed of 15 references, with abstracts, published by WHO.

BIBLIOGRAPHY

Brodeur, Paul. *Outrageous Conduct: The Asbestos Industry*. Pantheon, New York, NY, 1985.

Brookins, D. G. *The Indoor Radon Problem*. Columbia University Press, New York, NY, 1990.

Browner, C. "Where There's Smoke There's Lung Cancer," *The Wall Street Journal*, May 6, 1994.

Chiras, D.D. *Environmental Science, Action for a Sustainable Future*. Benjamin/Cummings Publishing, Redwood City, CA, 1994.

Eliopoulas, C., J. Klein, M. K. Phan, B. Knie, M. Gromwald, D. Chitoyat and G. Koran. "Hair Concentrations of Nicotine in Women and Their New-Born Infants," *Journal American Medical Association,* 271, Feb. 23, 1994.

Environmental Protection Agency. *A Citizen's Guide to Radon* (2nd edition). Washington, DC, 1992.

Environmental Protection Agency. *Radon Reduction Methods* (2nd edition). Washington, DC, 1987.

Hanson, S. *Managing Indoor Air Quality*. Fairmount Press, 1991.

H. R. 1530 - The Indoor Air Quality Act of 1989. Hearings July 20 and September 27, 1989. *101st Congress, Publication No. 89*, U. S. Superintendent of Documents.

Kerr, R. A. "Indoor Radon: The Deadliest Pollutant," *Science,* 39, April 29, 1988.

Miller, G. T., Jr. *Living in the Environment* (5th edition). Wadsworth Publishing Co., Belmont, CA, 1988.

Miller, G. T., Jr. *Resource Conservation and Management*. Wadsworth Publishing Co., Belmont, CA, 1990.

National Academy of Sciences. *Indoor Air Pollutants*. National Academy Press, Washington, DC, 1981.

Raven, P. H., L. R. Berg, and G. B. Johnson. *Environment*. Saunders College Publishing, Philadelphia, PA, 1995.

Sheldon, L. S. et al. *Indoor Air Quality in Public Buildings, Vol. I and II*, U. S. Environmental Protection Agency, Washington, DC, 1988.

Stone, R. "No Meeting of the Minds on Asbestos," *Science*, 254, November 15, 1991.

Turiel, I. *Indoor Air Quality and Human Health*. Stanford University Press, Stanford, CA, 1985.

DISCUSSION QUESTIONS

1. Should government attempt to regulate the permissible concentration of radon in a private home?

2. Do buildings on your campus contain asbestos? What areas would be most likely to contain significant amounts?

3. In what ways has energy conservation contributed to high levels of indoor air pollution?

4. Should we totally ban smoking in public places?

5. Is your home likely to have high concentrations of radon?

6. Does society have a duty to protect children from passive smoke from their parents?

7. Why are only certain individuals normally affected by "sick building syndrome"?

NOTES

1. EPA, Report to Congress on Indoor Air Quality, August, 1989.

2. Radon is directly regulated by the Indoor Radon Abatement Act (IRAA), 15 USC Sec. 2661 *et seq.*, although sections of other federal statutes, such as the Clean Air Act, could arguably be used, as discussed *infra*.

3. Asbestos is regulated by the Resource Conservation Recovery Act (RCRA), 42 USC Sec.6901 *et seq.*, the Comprehensive Environmental Response, Compensation and Liability Act (CERCLA), 42 USC Sec. 9601 *et seq.*, the Toxic Substance Control Act (TSCA), 15 USC Sec. 2601 *et seq.*, the Clean Air Act (CAA), 42 USC Sec. 7401 *et seq.*, the Occupational Safety and Health Act (OSHA), 29 USC Sec. 651 *et seq.*, and the Asbestos Hazard Emergency Response Act of 1986 (AHERA), 15 USC Sec. 2641 *et seq.*

4. Chlorofluorocarbons are regulated by the Clean Air Act, 42 USC 7401 *et seq.*, and its various amendments.

5. Over 90% of the states have enacted legislation designed to ban or restrict cigarette smoking in public places. Loewy, Steven A., *et al*, "Indoor Air Pollution in Commercial Buildings: Legal Requirements and Emerging Trends," 11 Temple Env. L. & Tech. J. 239, 252 (1992).

6. S.REP.NO. 304, 101st Congr., 2d Sess. 3 (1990).

7. Senate Bill 656, Cong. Rec. March 25, 1993, S3773 - 3784.

8. Kelly, Steve, "Indoor Air Pollution: An Impetus for Environmental Regulation Indoors," B.Y.U.J. Pub.L., Vol. 6:295 (1992).

9. Only California can be said to have a truly comprehensive legislative program.

10. For example, the American Society of Heating, Refrigeration and Air-Conditioning Engineers (ASHRAE)and the American Institute of Architects (AIA) have attempted to provide such guidance.

11. 42 USC 7401 *et seq.*, as amended by the Clean Air Act Amendments of 1990, Pub.L.No. 101-549 (November 15, 1990).

12. 42 USC 7602(g).

13. 40 CFR 50.1(c)(1991).

14. 15 USC 2601 *et seq.*

15. 15 USC 2605 *et seq.*

16. 29 USC 651 *et seq.*

17. 15 USC 2051 *et seq.*

18. 15 USC 2052(a)(1).

19. Kelly, *supra*, at 314.

20. 42 USC 9601, *et seq.*, as amended by the Superfund Amendment and Reauthorization Act, 42 USC 9659, *et seq.*

21. See, for example, *T&E Industries v. Safety Light Corp.*, 680 F.Supp. 696 (D.N.J. 1988), *Vermont v. Staco, Inc.*, 684 F.Supp. 822 (D.Vt. 1988).

22. Kelly, *supra*, at 316.

5
The Greenhouse Effect and Stratospheric Ozone Depletion

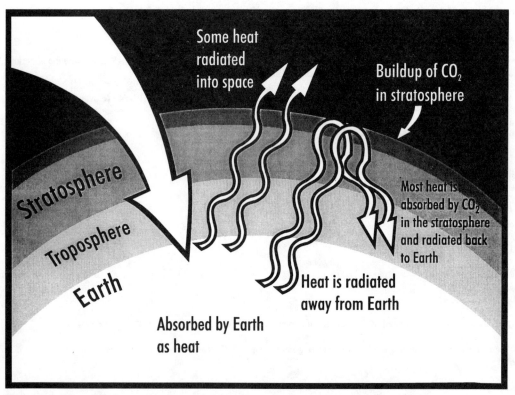

Figure 5-1: How global warming is produced by the greenhouse effect and carbon dioxide.

The Greenhouse Effect

The **greenhouse effect** is another name for global warming caused by carbon dioxide and other gases in the atmosphere. Sunlight passes through these gases to reach the surface of the Earth. The Earth then re-radiates this energy, but at longer wavelengths of the electromagnetic spectrum than visible light. Most of this radiation is in the infrared part of the spectrum. However, infrared radiation cannot pass through the "greenhouse gases." As a result, most of the heat radiated from the Earth is absorbed by the greenhouse gases and re-radiated in all directions. Only part of the re-radiated energy is directed into outer space; the remainder is directed back to the Earth. This leads to an increase in the temperature of the surface of the entire planet. This effect is similar to what happens in a greenhouse. Sunlight passes through the glass windows of the greenhouse. The infrared energy reradiated from the inside of the greenhouse cannot easily pass through the glass: this energy is trapped inside the greenhouse, thereby raising its temperature.

Figure 5-1 depicts the greenhouse effect. The most important greenhouse gas is carbon dioxide, a molecule consisting of one atom of carbon and two atoms of oxygen. The concentration of atmospheric carbon dioxide has increased from about 274 parts per million in pre-industrial times to 356 parts per million in 1992. By comparison, the next most abundant greenhouse effect gas, methane, has increased from 0.7 parts per million in preindustrial times to 1.7 parts per million in 1992. Other trace greenhouse gases such as chlorofluorocarbons (CFCs), nitrous oxide and tropospheric ozone, which have also increased, are presently at 0.3 parts per million or less.

Sources of Greenhouse Gases

Carbon dioxide is produced in the combustion of fossil fuels such as coal and oil. Factories, power-generating plants, and motor vehicles all contribute to the production of this greenhouse gas. Deforestation, particularly the wide-spread clearing and burning of tropical rain forests, also contributes to the increase in carbon dioxide in the atmosphere because there are fewer plants and trees to absorb carbon dioxide from the atmosphere during photosynthesis. Methane is produced in the digestive tracks of cattle, in wetlands, and in rice paddies. Nitrous oxide comes from the fertilization of soils and the concentration of animal wastes in feedlots. CFCs, which are the major cause of upper-air ozone depletion, come from air conditioners, aerosol cans, refrigerants, and other sources.

Effects of Global Warming

There is no doubt that greenhouse gases will contribute to the warming of the planet. That issue is not debated among scientists. What is hotly debated is the magnitude of the effect. The debate is rooted in the fact that other processes contribute to the average temperature of the surface of the Earth. Over the last century the average temperature of the planet has increased by about one degree Fahrenheit. However, between about 1940 and 1975 no warming occurred. Rather, half of the warming of the planet occurred before 1940, even though most of the emissions of greenhouse gases occurred after that date. One possible way to reconcile these observations with the greenhouse theory is to note that the planet has undergone cyclical temperature changes in the past. One geologist asserted in 1975 that despite an increase in the concentration of greenhouse gases the planet cooled somewhat during the 1960s and 1970s because it was at a low point in a natural 80-year temperature cycle. The geologist predicted the warming of the 1980s, when the natural temperature cycle turned upward, thereby adding to the warming effects of the greenhouse gases. In fact, the 1980s included 8 of the warmest years on record, with 1990 being the hottest. In 1993, records for the most number of hot days ever in a summer were set in many locations in the United States. Figure 5-2 illustrates the result of the natural temperature cycle superimposed on the warming trend caused by the steadily increasing concentration of carbon dioxide in the atmosphere.

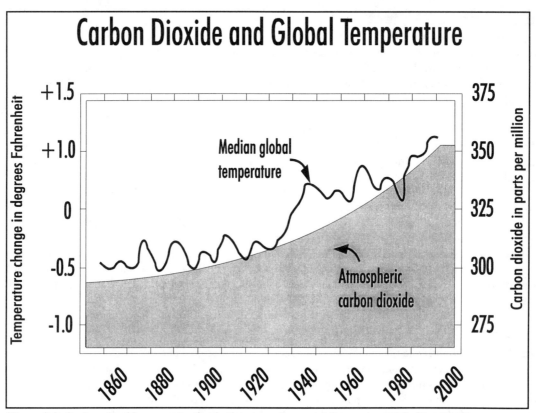

Figure 5-2: Carbon dioxide concentration in the atmosphere and the average global temperature.

Among the other processes which can mask the effect of the greenhouse effect is sulphur haze, a kind of air pollution which causes acid depositions on the planet. It also cools the planet by reflecting sunlight back into outer space before it reaches ground level. Sulphur emissions come from the same factory smokestacks that emit carbon dioxide. In addition, volcanic eruptions can send huge amounts of sulphur-containing particles into the atmosphere. The most powerful volcanic eruption this century—at Mount Pinatubo, in the Philippines, June 1991—hurled huge amounts of sulphur into the stratosphere. Sulphur that reaches the stratosphere tends to remain for a much longer time than when the particles are confined to the lower troposphere. The effect of this eruption may cause global cooling of about one degree Fahrenheit for a period of several years. This would obviously help to mask greenhouse warming. In the long run, greenhouse warming will probably overcome sulphur cooling because the greenhouse gases will remain in the atmosphere for hundreds of years, whereas human-produced sulphur emissions (which remain at lower altitudes) will fall to Earth in days or weeks. Furthermore, the greenhouse gases warm the planet for 24 hours a day by re-radiating the energy being radiated out into space while the sulphur acts to cool the Earth only during daylight hours.

Climatologists have developed models which operate on computers to predict the effects of further increases in greenhouse gases. However, these models may not adequately represent the complicated interactions between the atmosphere, land, and oceans. Current models predict that a doubling of the carbon dioxide content in the atmosphere will cause the average temperature of the Earth to increase by 4 to 9 degrees Fahrenheit.

An increase of several degrees Fahrenheit in global temperature would have dramatic effects. While lower heating bills might occur in middle and higher latitudes, and crop yields might increase in some areas due to both a longer growing season and an increase of carbon dioxide in the atmosphere (enhancing the rate of photosynthesis), the negative impacts would be catastrophic. Warmer temperatures would cause water to expand and also would lead to partial melting of mountain glaciers and the polar icecaps. These effects would raise global sea levels by perhaps two feet. This would flood large areas of agricultural lowlands and deltas in China, India, and Bangladesh, causing widespread famine. Large parts of Louisiana and Florida would end up under water. The net result would be to disrupt the world's food supply for many years, resulting in economic disasters and sharply higher food prices. Huge investments in fertilizers, irrigation, and dike systems would be required, costing hundreds of billions of dollars.

The worst-case scenario is one in which a number of positive planetary feedbacks are caused by greenhouse warming. A feedback is a component of the world-wide climate system that is triggered by other components of the system. A **positive feedback mechanism** is one that would amplify greenhouse warming, in contrast to negative components, which would reduce the warming. Consider the following scenario: as the world warms, the oceans are less able to absorb carbon dioxide since carbon dioxide is less soluble in warmer water. Then follows a drop in the creation of biomass by the phytoplankton in the ocean. This is caused by a reduced deep-ocean nutrient supply to the upper levels of the oceans, which are more stable than usual due to the effects of warming. In addition, there are increases in ultraviolet radiation (see the next section on ozone depletion in the stratosphere) which reaches the ocean above the highly productive waters of the Antarctic and sub-Arctic. This further decreases the amount of phytoplankton in the ocean and thus further decreases the absorption of carbon dioxide by the oceans. As temperatures continue to rise, the tundra becomes a vast wetlands releasing enormous amounts of methane, an even more potent greenhouse gas than carbon dioxide. Next, more high altitude clouds form, containing ice crystals which help to trap even more heat in the atmosphere.

The point is not that any of these positive feedback mechanisms *will* occur but rather that many scientists think that each is possible in a warming environment. In the event that this worst-case scenario did occur it would be too late for the human race: the warming could not be reversed by any actions we could take.

IPPC Report on Global Warming
In 1988 the United Nations Environment Program and the World Meteorological Organization established the Intergovernmental Panel on Climate Change (IPCC) to provide periodic comprehensive assessments on climate change to guide policy makers internationally. In Fall 1995, with input from over 2,000 scientists, the IPPC made a Second Assessment Report which included several key findings. They are:

- Human activities are increasing the atmospheric content of carbon dioxide and other greenhouse gases that tend to warm the atmosphere.

- The surface temperature of the Earth has increased by 0.5 to 1 degree Fahrenheit over the last century.

- The balance of evidence suggests that there is a measurable human influence on global climate.

- Models that take account of the observed increases in the atmospheric concentrations of greenhouse gases are simulating the recent history of observed changes in surface temperature and its vertical distribution with increasing realism.

- Unless there is a reduction in the growth of greenhouse gas emissions, the Earth's average temperature is expected to increase by about 2 to 6.5 degrees Fahrenheit by the year 2100. The sea level will rise by 6 to 38 inches by the year 2100.

Other scientists took exception to these predictions. One point raised was that satellite data, available since 1979, shows no global warming but rather a slight cooling. The response to this was that satellite data measures the temperature several miles above the Earth. At ground level the temperature has indisputably increased. A second major objection is that the computer models used in these projections don't adequately account for the most important greenhouse gas, water vapor. One critic asserted that a doubling of carbon dioxide concentrations in the atmosphere would be equivalent to a change in average relative humidity of only 4%. The scientists who wrote the IPCC Report concede the limitations of their models. That is why there is such a broad range of possible temperature and water level increases predicted.

Methods of Diminishing the Effects of Greenhouse Gases

There are two general approaches to the problem of diminishing the effects of greenhouse gases. The first attempts to reduce the emissions of greenhouse gases. The second approach adjusts our activities to compensate for greenhouse warming.

One obvious action is to reduce the use of fossil fuels. This could be achieved by some combination of energy conservation, increased use of nuclear power, and increased use of renewable power sources such as solar energy or wind-driven electric generators. Planting forests world-wide would also help because the additional vegetation would absorb more carbon dioxide from the atmosphere. However, there are major obstacles to overcome in following any of these courses of action. It is unlikely that most countries will reduce their consumption of fossil fuels or limit deforestation because there is too great a divergence among the scientific community as to the pace and magnitude of global warming. Governments are more likely to believe best-case scenarios, especially since implementing significant reductions in greenhouse gas emissions are sure to cause major short-term economic and social disruptions which most governments would find politically too unpalatable to implement.

Many experts believe that we should prepare for long-term global warming by developing crops that need less water or that can thrive on water that is too salty for present crops. Large-scale dike systems need to be constructed to reduce flooding in coastal areas. Building in low-lying regions should be curtailed. Food stockpiles should be increased world-wide.

One science fiction-like solution is to construct a large number of screens to be put into orbit between the sun and the Earth in order to block out part of the sunlight (thereby counteracting greenhouse effect warming). The screens would also intercept part of the ultraviolet radiation (thereby compensating for the depletion of ozone in the stratosphere).

What Can You Do to Prevent Global Warming?

The Environmental Protection Agency has a list of the top ten things anyone can do to help prevent global warming. The actions and the average per family reductions of carbon dioxide emissions in the atmosphere are as follows:

1. Purchase a fuel-efficient car (rated at 32 miles per gallon or better).
   Carbon dioxide reduction =5,600 lbs/year.

2. Insulate your home, get a more efficient furnace, and install energy-efficient showerheads.
   Carbon dioxide reduction = 2,480 lbs/year.

3. Leave your car at home two days a week.
   Carbon dioxide reduction = 1,590 lbs/year.

4. Recycle all of your home's waste newsprint, cardboard, glass and metal.
   Carbon dioxide reduction = 850 lbs/year.

5. Install a thermal system to provide your hot water.
   Carbon dioxide reduction = 720 lbs/year.

6. Replace your current washing machine with a low-energy low-water-use machine.
   Carbon dioxide reduction = 440 lbs/year.

7. Buy products in reusable or recyclable packaging.
   Carbon dioxide reduction = 230 lbs/year.

8. Replace your home refrigerator with a high-efficiency model.
   Carbon dioxide reduction = 220 lbs/year.

9. Use a push lawn mower rather than a power mower.
   Carbon dioxide reduction = 80 lbs/year.

10. Plant a couple of trees around your home.
    Carbon dioxide reduction = 20 lbs/year.

Ozone Layer Depletion

Ozone is a molecule consisting of three oxygen atoms. It is a human-made pollutant. In the lower atmosphere it is harmful to human health and that of plants and animals. In the stratosphere (6 to 28 miles above the Earth's surface) the ozone provides a shield which screens out more than 99% of the sun's harmful ultraviolet radiation. Figure 5-3 illustrates the way ozone in the stratosphere absorbs ultraviolet radiation from the sun and prevents most of it from reaching the Earth. Should ozone ever disappear from the stratosphere, the Earth would become uninhabitable for many forms of life. In the summer of 1990, NASA reported that globally the ozone layer had been depleted by 2 -3% over the previous two decades. It was also reported that the ozone layer had already begun to thin over the United States and other populated areas in the middle latitudes.

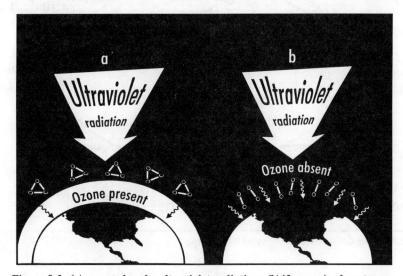

Figure 5-3: *(a)* **ozone absorbs ultraviolet radiation;** *(b)* **if ozone is absent more ultraviolet radiation reaches the Earth.**

The primary cause of stratospheric ozone-layer depletion is the use of chlorofluorocarbons (CFCs). CFCs were first utilized in the 1930s. They were found to be non-toxic to humans, inexpensive to produce, and energy-efficient substitutes for ammonia (which was potentially explosive) as the coolant in refrigerators. CFCs were marketed under the trade name Freon and soon became the leading coolant for air conditioners and refrigerators. They also served as propellants for spray cans, which were used in the 1960s and early 1970s to apply everything from hair sprays to deodorants to household cleaners. Plastic manufacturers incorporated CFCs to make foam products such as seat cushions, packing materials, and insulation. More recently, the electronics industry found that the chemical stability of CFCs made them ideal cleaning solvents for components such as silicon chips and circuit boards.

In 1960, CFC production was about 150,000 tons per year. By 1974, it had grown to over 800,000 tons per year. Furthermore, as CFCs became more and more widely utilized for industrial applications, other chemicals that also were later found to deplete the ozone layer were coming into use. Two other chlorinated compounds, methyl chloroform and carbon tetrachloride, became widely used for cleaning and degreasing metal parts in industrial processes and as ingredients in pesticides. Halons were developed for use in fire extinguishers.

In 1974, researchers investigated the possible effects of chlorine released by NASA rockets on stratospheric ozone. They concluded that a single atom of chlorine would destroy many thousands of ozone molecules. Later that year other researchers conducted a study of CFCs released into the atmosphere. They concluded that CFCs are so durable that they float around in the air in their original state for many years, eventually drifting upward into the stratosphere where they release chlorine atoms that attack ozone. Figure 5-4 presents the cycle of destruction of ozone by chlorine. A free chlorine atom reacts with a molecule of ozone, creating an oxygen molecule, and a molecule of chlorine oxide. The chlorine oxide then reacts with a second molecule of ozone to form two molecules of oxygen and a free chlorine atom which is then available to repeat the cycle. For every two ozone molecules that chlorine destroys, three oxygen molecules are produced. This process destroys ozone at a faster rate than that at which replacement ozone molecules are being created by natural processes.

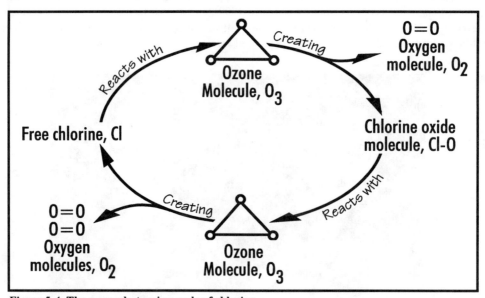

Figure 5-4: **The ozone-destroying cycle of chlorine**

The debate over the amount of ozone destruction from CFCs continued for some years. In 1978, the United States became the first country to ban the use of CFCs as propellants in aerosol spray cans. In 1985, a team of British scientists discovered the existence of an ozone "hole," or thin spot, in the stratosphere above Antarctica where ozone levels decrease by as much as 67% each year. The hole in the ozone layer over Antarctica occurs annually between September and November, when the circumpolar vortex, a mass of cold air, circulates around the southern polar region, in effect isolating it from the warmer air in the rest of the world. This cold causes the formation of stratospheric clouds containing ice crystals to which chlorine adheres. The chlorine then attacks ozone. When the circumpolar vortex breaks up each year, the ozone-depleted air spreads northward, decreasing ozone levels in the stratosphere over South America, New Zealand, and Australia. A smaller ozone hole has been detected in the stratosphere over the Arctic. In 1995, researchers Paul Crutzen, Mario Molina, and Sherwood Rowland shared the Nobel Prize in Chemistry for their contributions to understanding how ozone is formed and decomposes in the atmosphere.

Natural phenomena may also deplete the ozone layer. Studies of the impact of volcanic ash in the aftermath of the 1982 eruption of El Chichon in Mexico found that ozone concentrations over the middle latitudes were significantly depleted.

Table 5-1 presents a summary of ozone-destroying chemicals in the atmosphere and gives an estimated lifetime in the atmosphere for each chemical. As one can see, CFCs represent 80% of the problem. The U.S. share of the world-wide use of each chemical is also listed. (Source: US Office of Technology Assessment)

Table 5-1: Ozone-Destroying Chemicals in the Atmosphere

Chemical	Uses	Lifetime in atmosphere	Share of problem	U.S. share of use
CFC-11	coolant, aerosol, foam	60 yr	28%	22%
CFC-12	coolant, aerosol, foam	13 yr	47%	30%
CFC-113	solvent	90 yr	5%	45%
Carbon tetrachloride	solvent	50 yr	15%	27%
Methyl chloroform	solvent	7 yr	2%	50%
Halon 1211	coolant, foam	25 yr	1%	25%
Halon 1301	fire extinguisher	110 yr	2%	50%

Effects of Ozone Layer Depletion

There are many dangers from increased ultraviolet radiation reaching the surface of the Earth. Sun lovers should take note: even the most powerful sunscreens may not block out the damage to the immune system from ultraviolet radiation. Exposure can also lead to accelerated aging of the skin and various forms of skin cancer. In addition, increased cataract development is possible. The quantitative estimate of the extent of human health damage is dependent upon the amount of ozone depletion on a world-wide scale. If ozone erosion continues at perhaps 4% per decade, year-round and world-wide, one report estimates that we could experience an extra 12 million skin cancer cases in the United States over the next 50 years.

Radiation also slows the growth of phytoplankton, which are the mainstay of the ocean food chain. Interference with photosynthesis in plants could result in lower crop yields. The severity of the consequences of stratospheric ozone depletion may also depend on latitude. Ozone losses at the equator may have a greater biological effect than the same losses at a higher latitude.

Some Myths about Stratospheric Ozone Depletion

Since CFCs are heavier than air, some have claimed that these molecules could not rise up to the stratosphere. The fallacy of this argument is that while CFCs are heavier than air, and will accumulate near the floor in a still room, the atmosphere is in constant motion. It is this motion, due to winds, that mixes CFCs nearly uniformly worldwide.

A second myth is that volcanoes and the oceans are causing stratospheric ozone depletion, not CFCs. It is true that volcanic eruptions do contain hydrogen chloride and that the oceans do produce large amounts of sea salt, which contains chlorine. However, most volcanic eruptions are too weak to reach the stratosphere. Sea salt from the oceans is also released very low in the atmosphere. In addition, both sea salt and hydrogen chloride are highly soluble in water, unlike CFCs which do not dissolve in water. Hence rain effectively "scrubs" the lower atmosphere of both sea salt and hydrogen chloride.

A third myth is that stratospheric ozone depletion occurs only in Antarctica. In fact, stratospheric ozone levels in the middle latitudes, in which most of the world's population live, have fallen by 10% during the winter and 5% during the summer months over the time period 1979-1994.

Methods of Diminishing the Use of Ozone-Destroying Chemicals

The immediate solution to the use of CFCs as coolants will be to replace them with chemicals less damaging to stratospheric ozone. One of the more intriguing long-term solutions for phasing-out CFCs as refrigerants is to cool using sound waves rather than chemicals. Basically, a thermoacoustic refrigerator uses a loudspeaker at the end of a tube which emits an extremely loud noise that resonates inside the gas-filled tube to form a "standing wave" in which the individual gas molecules are rapidly oscillating while the wave as a whole does not move. The gas molecules move back and forth at the same frequency as the loudspeaker and carry heat away from the thinner regions of the standing wave towards its thickest bulge, called its antinode. If a metal plate is placed within this antinode the plate absorbs heat from the gas molecules and conducts the heat away from the tube, leaving cool gas in its place. The cool gas, in turn, withdraws heat from the space to be cooled. There are many technological problems to be overcome before such refrigerators are economically viable. For example, the sound produced by the loudspeaker is so intense that it would destroy living tissue exposed to it. Thus, safety considerations become important in the construction of such a new type of refrigerator.

In the electronics industry, CFCs used to clean solder flux from circuit boards have already been replaced by a number of simple procedures such as dunking the circuit boards in soapy water, shaking them, and blow-drying them with hot air.

What Can You Do to Prevent Stratospheric Ozone Depletion?

The Environmental Protection Agency recommends:

1. Make sure that technicians working on your car or home air conditioner or refrigerator prevent the refrigerant from being released into the atmosphere (this is the law).

2. Have your car and home air conditioning units checked for leaks.

3. Properly dispose of refrigeration or air conditioning units (this is also the law).

SOCIETY'S REACTIONS TO THESE ENVIRONMENTAL ISSUES

Once society learned of scientific evidence that a buildup of greenhouse gases was contributing to the overall warming of the Earth's surface, and that stratospheric ozone was being depleted, the issue of how to mitigate these phenomena arose. Voluntary efforts could not be counted on to solve the problems, so like other issues in society, the greenhouse/ozone problems began to receive attention from state and federal governments in the United States (as well as in other countries throughout the world). This attention took the form of legislation, which represents government's primary method for controlling or reducing an element which government has determined is undesirable. Unlike other issues in society, however, government could not simply mandate a direct solution to the problem; in other words, government could not, for example, order the ozone layer to stop depleting. Instead, government had to look at the causes of the ozone depletion/greenhouse problem and enact legislation which had an impact on the sources.

As discussed earlier in this chapter, the use of chlorofluorocarbons (primarily as cooling agents), the burning of fossil fuels, and deforestation are the primary causes of the ozone depletion/greenhouse problems. While the severity of the ozone depletion/greenhouse problems was apparent, governments could not prohibit all fossil fuel and chlorofluorocarbon use without drastically affecting life as we know it and causing economic disaster. Instead, the approach has been focused on minimizing the effects of fossil fuel use and other air emissions, promoting alternative energy sources, requiring the phasing out of chlorofluorocarbon use (once a replacement compound had been found), and placing limits on the large-scale cutting of trees.

No single type of legislation could have the multiple effects described in the previous paragraph. Instead, different types of legislation produce different results. In essence, there are three types of legislation which are intended to produce the effects discussed, and these types can be characterized as: 1) control-based legislation; 2) incentive-based legislation; and 3) process-oriented legislation.[1]

Control-Based Legislation

Control-based legislation includes those laws and regulations which contain government-mandated goals, the attainment of which is enforced through the imposition of sanctions for non-compliance. Most environmental legislation in general constitutes this type, and specific environmental statutes which address the ozone depletion/greenhouse problems include the federal Water Pollution Control Act, known as the Clean Water Act of 1977,[2] parts of the Clean Air Act,[3] the Solid Waste Disposal Act,[4] the Resource Conservation and Recovery Act,[5] and the Toxic Substances Control Act.[6] These statutes represent the most common form of environmental legislation, since the most basic approach to prohibiting detrimental impacts on the environment is to place limits on the detrimental activity. Limits on the form and extent of pollutant discharges to the water and air and limits on the type and place of disposal, handling, use, and generation of toxic substances are enforced through the imposition of fines and penalties which can be substantial.

Further underscoring the seriousness of these laws is the extent of liability which can be imposed. Most environmental statutes impose liability for violation of established limits regardless of whether the violator knew of the existence of the law or intended for the violation to occur. This "strict" liability arises even though the violator was not negligent in its course of conduct. In certain situations, criminal penalties may be levied where violation was knowing or reckless. Strict liability, while not unique to environmental law, gives environmental, control-based legislation the force necessary to ensure industry's respect of statutorily mandated limits.

Although all control-based legislation imposes limits on polluting activity, the limits mandated by statutes like the Clean Air Act and the Clean Water Act force industry to create and adopt new technologies which are more efficient at reducing those pollutant discharges which directly or indirectly impact the amount of pollutants in the air that contribute to the ozone depletion/greenhouse problems. Although the medium may be different (that is, water versus air), the goal is the same: reduce pollution being released into the environment. Clearly, however, the Clean Air Act has the most direct impact on those processes which result in ozone depletion and greenhouse-effect enhancement.

Although all environmental problems have an impact, to a varying degree, outside of the border of the country in which the problems originate, the ozone depletion/greenhouse effects are truly world-wide. Realization of the severity of the situation and implementation of corrective action by all industrialized nations is essential if the problems are to be corrected. It is logical, therefore, that the ozone depletion/greenhouse problems would be the impetus for the first global meetings on an environmental issue. In 1987, the United States and 23 other countries signed the Montreal Protocol on Substances That Deplete the Ozone Layer. These countries promised to reduce their production and importation of CFCs and halons by 50% no later than 1999. As further research indicated an ozone loss more severe than previously thought, in 1990 the signatories to the Montreal Protocol revised the treaty to require the total elimination of CFC production and importation by the year 2000 and significantly expanded the list of compounds targeted for phase out. The resulting Clean-Air Act Amendments became U.S. law in November 1990.

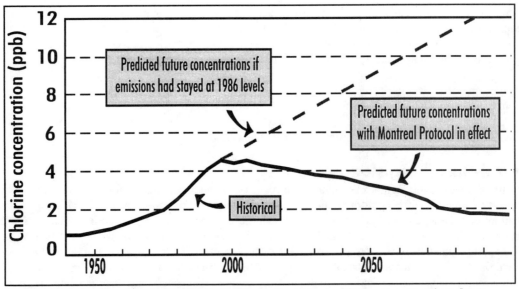

Figure 5-5: **Estimation of the anticipated reduction in atmospheric chlorine concentrations due to compliance with the Montreal Protocol.**

In November, 1992, the Montreal Protocol was again revised to ban the production of CFCs beginning January 1, 1996. Hydrochlorofluorocarbons (HCFCs) will be restricted beginning in 2010. HCFCs have been introduced as a temporary substitute for CFCs used in refrigeration and air conditioning systems for big buildings. Although they are chemically similar to CFCs, these chemicals are less stable and break down more readily when they are released into the atmosphere. For that reason, HCFCs are far less damaging to the ozone layer. Hydrofluorocarbons (HFCs) are also a transitional substitute for CFCs in mobile air conditioners and home refrigerators. The Montreal Protocol also requires recycling to reduce the release into the atmosphere of existing stores of ozone-depleting chemicals. Responsibility for implementation of the Protocol and other mandates which resulted from these meetings, however, remains with the individual countries and no overall enforcing authority was created. Figure 5-5 presents the overall anticipated decrease in chlorine concentrations (in parts per billion) in the atmosphere due to the implementation of the Montreal Protocol. Note that atmospheric chlorine concentrations are not projected to return to 1975 levels until the year 2075.

Even before the Clean Air Act Amendments of 1990 were enacted, Title VI of the Clean Air Act dictated that a phase out of ozone-depleting substances begin by 1988. The scope and impact of the 1990 Amendments, however, far exceed that encompassed by prior amendments to the Act. The 1990 Amendments require many states to substantially revise their own clean air legislation and regulatory provisions. One target of such revisions are automobile emissions. In addition to requiring states to implement stringent controls and procedures to reduce emissions, those states within a "severe nonattainment area" must develop employer trip reduction programs to reduce traffic congestion and miles traveled by single-occupant vehicles commuting to work.[7] The amendments also require states to implement operating permit programs to ensure consistent and more efficient regulation of industrial emissions[8] and other air pollution control measures.

Incentive-Based Legislation

This type of legislation attempts to encourage certain activities to occur and certain goals to be met through the creation of economic incentives, which take the form of tax credits, allowable tax deductions, and other forms of credits. Likewise, negative incentives may also be implemented, which may involve a tax on activities or substances connected with causing pollution. While incentive-based legislation is used often in an economic context,[9] the application of this concept to environmental issues is relatively new. One piece of economic legislation contained a provision which actually "crossed over" into environmental law. The Omnibus Budget Reconciliation Act of 1990 contained a provision which imposed an excise tax on ozone-depleting substances to create disincentives to the continued use of such substances in the marketplace.[10] As another example, the Clean Air Act Amendments of 1990[11] contain provisions which give industry the ability to buy and sell "pollution credits" to foster attainment of stringent air-quality criteria for a given region at less economic cost than if such buying were not allowed. While the Act includes more stringent standards than those previously in place, there is an implementation schedule to allow industry and consumers to become accustomed to the new requirements. There are also incentives for early attainment of the Act's goals. Although the impact of many of the amendments has yet to be felt, the Clean Air Act's incentive-based provisions may prove to be the most successful approach to regulating pollution.

Process-Oriented Legislation

Unlike the types of legislation described above, statutes which are "process-oriented" do not seek to protect a particular resource, regulate a particular substance or activity, or create incentives. Instead, this type of legislation creates a framework within which a particular activity may be evaluated for the harm it may cause to society and a decision made as to whether to permit the activity to occur. This legislation is embodied by the

National Environmental Policy Act of 1969 (NEPA).[12] It requires that each federal agency charged with administering legislation must consider the environmental impact of federal legislation and prepare, prior to enactment, an Environmental Impact Statement (EIS) which describes:

1. the environmental impact of the proposed action,
2. any adverse environmental effects which cannot be avoided should the proposal be implemented,
3. alternatives to the proposed action,
4. the relationship between local short-term uses of the environment and the maintenance and enhancement of long term productivity, and
5. any irreversible and irretrievable commitment of resources which would be involved in the proposed action should it be implemented.[13]

The requirement to undertake a consideration of the environmental impact of proposed legislation has not, however, translated effectively into a substantive weighing of the benefits and detriments to the environment. In other words, the actual impact of NEPA has been a procedural one, rather than a substantive one. Providing that federal agencies prepare an EIS in conjunction with proposed legislation, the agency's decision to proceed with a specific piece of legislation which has a negative EIS is within the discretion of the agency. NEPA does not include provisions which require the agency to reduce the detrimental impact to the environment. Without such provisions, a decision of a federal agency to proceed with legislation is difficult to challenge.[14]

There has been legislation introduced which would amend NEPA to include provisions requiring that the federal government consider the effects of government action on global warming, but no such legislation has, to date, made it past congressional committee consideration.[15]

Role of the States

Although not discussed here in detail, individual states have a wide range of discretion to enact their own legislation of the type described, providing that the federal government has not reserved unto itself the exclusive authority over specific subjects.[16] The states may not, however, enact legislation which is more lenient than, or whose goals conflict with, the similar federal statute; that is, the state statutes must be at least as stringent as the federal statute. In many situations, the federal government may delegate the authority to enforce the federal statute to particular states which have met certain criteria outlined in those statutes. Such criteria may include the enactment by the state of legislation which "adopts" the federal version of the statute at a minimum.

Thus, federal control-based, incentive-based and process-oriented legislation is not the only (and may not even be the most severe) legislation which impacts on the sources of the ozone depletion/greenhouse problems. The concern over the seriousness of the problems translates into various degrees of environmental activism at the state and federal levels.

RELEVANT EDUCATIONAL AUDIO-VISUAL MATERIALS

Note: In some cases, A/V materials can be previewed before purchasing or renting.

The Greenhouse Effect. (1990) 18-minute color videotape. Available from: Media Design Associates, Inc., Box 3189 Boulder, CO 80307-3189, (800) 228-8854 / FAX (303) 443-2882.

Hothouse Planet. Software program for PCs and Macs. Available from: EMC Corporation, 41 Kenosia Avenue, P.O. Box 2805, Danbury, CT 06813-2805.

Global Warming - Future Quest. (1993) 31-minute color videotape. Available from: Hawkhill Associates, Inc., 125 E. Gilman Street, P.O. Box 1029, Madison, WI, 53701-1029.

Our Ozone Crisis. Two-part software program for PCs or Macs. *Part 1: Reduction in the Ozone Shield; Part 2: Effects of Volcanic Eruptions.* Available from: EMC Corporation, 41 Kenosia Avenue, P.O. Box 2805, Danbury, CT 06813-2805.

The Greenhouse Effect. (1988) 23-minute videotape. Available from: Public Media, Inc. (Films, Inc.), 5547 N. Ravenswood Avenue, Chicago, IL 60640-1199, (312) 878-2600 / (800) 826-3456 / FAX (312) 878-8406.

Greenhouse Effect: To What Degree? (1989) 22-minute videotape. Available from New Dimension Media, Inc. 85803 Lorane Highway, Eugene, OR 97405 (503) 484-7125 / (800) 288-4456 / FAX (503) 484-5267.

The Infinite Voyage: Crisis in the Atmosphere. (1989) 60-minute color documentary film produced by public television station WQED as part of the series, *The Infinite Voyage*, in collaboration with the National Academy of Sciences. It describes the growth of greenhouse gases in the atmosphere and the depletion of ozone in the stratosphere. Available from: Movies Unlimited, 6738 Castor Ave, Philadelphia, PA 19149 (215) 722-8298.

INTERNET RESOURCES
Access the following sites for further information. (Note: URLs are subject to change)

EPA Site for Global Warming
http://www.epa.gov/globalwarming/home.htm

This Environmental Protection Agency site is a source of a great deal of information about global warming, including recent research reports and legislative actions.

EPA Site for Stratospheric Ozone Depletion
http://earth1.epa.gov:80/docs/ozone

This Environmental Protection Agency Site is a source of a great deal of information about stratospheric ozone depletion, including recent research reports and legislative actions.

BIBLIOGRAPHY
Allen, John L., comp. *Annual Editions: Environment.* 1993/94.

"Attacks on IPPC Report Heat Controversy Over Global Warming," *Physics Today*, 55, August, 1996.

"Cooling with Sound: An Effort to Save Ozone Shield," *The New York Times*, February 25, 1992.
"Giant Screen Might Save the Earth," *USA Today* 121: 2577, June 1993.

"Global Warming: the Controversy Heats Up," *Current Health* 2 20: 5, January 1994.

"Good Science, Weird Reporting," *E 4:* 6, December 1993.

Hobson, A. "Ozone and Interdisciplinary Science Teaching—Learning to Address the Things that Count Most," *Journal of College Science Teaching*, September/October 1993, p. 33.

Miller, G. T., Jr. *Living in the Environment,* 5th Ed. Wadsworth Publishing Company, Belmont, California, 1988.

"Ozone Depletion and the Immune System," *The Lancet* 342: 8880, November 6, 1993.

"Ozone Depletion," *CQ Researcher* 2: 13, April 3, 1992.

"Producers of CFC Alternatives Gear Up for 1996 Phaseout," *Chemical and Engineering News,* July 4, 1994.

Raven, P. H., L. R. Berg, G. B. Johnson. *Environment.* Saunders College Publishing, Philadelphia, 1995.

"Scientific Debate Continues to Rage Over Rising CO_2, Global Warming," *The Chronicle of Higher Education* 39: 47, July, 28. 1993.

"Stratospheric Ozone Depletion," *The Lancet* 342: 8880, November 6, 1993.

"Turning Up the Heat," *Consumer Reports* 38, September 1996.

"The Widening Ozone Hole," *The Chronicle of Higher Education* 40: 22, February 2, 1994.

DISCUSSION QUESTIONS

1. Why is carbon dioxide the most important greenhouse gas?

2. Why is the destruction of tropical rain forests relevant to the greenhouse effect?

3. Would an increase in average temperature of several degrees Fahrenheit affect the yield of some of the major food crops in the world?

4. Why would the use of electric cars have an impact on future greenhouse warming?

5. Research the suggestion of placing a large number of screens into orbit between the sun and the Earth.

6. Why did the use of CFCs rise so dramatically in the 1960s?

7. What are the major health hazards if world-wide ozone layer depletion occurs?

8. Consider other alternatives to the use of CFCs and HCFCs to provide cooling and long-term storage of foods.

9. Which poses a greater threat to our future, stratospheric ozone layer depletion or the greenhouse effect?

NOTES

1. Fernandez, Jose L., "Global Warming Legislation: Putting the Carbon Genie Back in the Bottle," *Syracuse Law Review*, Vol. 42:1095, 1991.

2. 33 USC 1251 *et seq.*

3. 42 USC 7401 *et seq.*, as amended by the Clean Air Act Amendments of 1990, Pub.L.No. 101-549 (November 15, 1990).

4. 42 USC 6901 *et seq.*

5. 42 USC 6941 *et seq.*

6. 15 USC 2601 *et seq.*

7. 42 USC 7511a(d)(1)(B). Only certain employers are affected: those employing 100 or more employees at any single work location in an affected area. Certain exemptions are available, as well.

8. 42 USC 7661a(d).

9. For example, to encourage re-development of blighted urban areas through the designation of those areas as Enterprise Zones for which businesses who locate within them receive tax-based incentives, grants, and low-interest or subsidized loans.

10. 26 USC 4682.

11. Pub.L.No. 101-549 (November 15, 1990)

12. 42 USC 4321 *et seq.*, Pub.L.No. 91-190, 83 Stat. 852 (1970)

13. 42 USC 4332(C).

14. See, for example, *Vermont Yankee Nuclear Power Corp. v. Natural Resources Defense Council, Inc.,* 435 US 519 (1978).

15. For example, the Global Environmental Research and Policy Act of 1989, H.R. 980, 101st Cong., First Session (1989) and the Global Warming Prevention Act of 1989, H.R. 1078, 101st Cong., First Session (1989).

16. Such as the federal government's sole authority over nuclear energy and nuclear power issues in the original Atomic Energy Act of 1946, Pub. L. No. 79-855, 60 Stat. 755.

6
Electromagnetic Waves

Table 6-1: The Electromagnetic Spectrum

Uses	Frequency	Spectral Frequency	Wavelength
Power Transmission	300 Hz	Extremely Low	1,000,000 m
			100,000 m
	30,000 Hz	Very Low	10,000 m
		Low	1,000 m
	3×10^6 Hz	Medium	100 m
Radio		High	10 m
Television	3×10^8 Hz	Very High	1 m
Radar		Ultra High	10^{-1} m
	3×10^{10} Hz	Super High	10^{-2} m
Microwaves		Extremely High	10^{-3} m
	3×10^{12} Hz		10^{-4} m
Radiating Heat		Infrared	10^{-5} m
	3×10^{14} Hz		10^{-6} m
Visible Light		Ultraviolet	10^{-7} m
Sun Lamps	3×10^{16} Hz		10^{-8} m
			10^{-9} m
	3×10^{18} Hz	X-rays	10^{-10} m
		Gamma Rays	10^{-11} m
	3×10^{20} Hz		10^{-12} m
		Cosmic Rays	10^{-13} m
	3×10^{22} Hz		10^{-14} m

Electromagnetic Radiation

Electromagnetic radiation surrounds us. Its spectrum (Table 6-1) extends from waves that barely vary (i.e. have an extremely low oscillation frequency and a long wavelength), to cosmic waves that vary greatly (i.e. have an extremely high frequency and a short wavelength). Between these extremes, there are particular frequencies or ranges of frequencies that are quite common in our environment because we have harnessed them for our own purposes or they are produced by the sun. These anthropogenic electromagnetic waves include: (1) 60 Hz, the frequency of electric power generation, transmission, and use, which is more properly separated into electric and magnetic fields; (2) the range of frequencies of electromagnetic waves used for commercial radio and television (in bands of frequencies from 535 kHz to 890 MHz) and for microwave ovens (915 and 2450 MHz).

Beyond these frequencies, and beyond visible light, is the ultraviolet portion of the spectrum. While there are man-made sources of exposure to ultraviolet, the predominant source is the sun.

These three portions of the electromagnetic spectrum are explored in this chapter. They are prevalent in the environment and they have distinctly different mechanisms of interacting with and affecting humans. Some of their reported effects are quite controversial, especially those suggested for the extremely low frequency, 60-Hz electric and magnetic fields, which we will explore first.

60-Hz Electric and Magnetic Fields

At low frequencies and long wavelengths, electromagnetic waves do not radiate as they do from a radio tower. They can be considered to be separate electric and magnetic fields that are quasi-static, or nearly stationary, and that behave independently. This means that their interaction and effects in the environment can be considered separately.

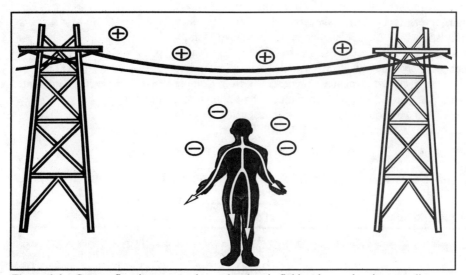

Figure 6-1a: Current flow in a person due to the electric field under an electric power line.

People **conduct** electricity (i.e. electric charges can move fairly easily through human tissue), so when they are in an **electric field**, charges in the body tend to move because of the force exerted by the field. The electric field produced by a 60-Hz electric power line changes direction 120 times each second, and this causes currents to flow back and forth in conducting objects, including people, near it. The direction of the current is the same as the electric field direction, and this is generally vertical near an electric line (Figure 6-1a). Also, because people conduct, the field does not penetrate far into the body, and the currents tend to be close to the surface. This is similar to the distribution of electric currents produced in a metallic object placed in an electric field.

Magnetic fields are produced by electric currents, and electric currents in power lines create magnetic fields that also cause current to flow in people. In this case, however, the current flows circumferentially, in loops around the body, and it is greatest near the perimeter (Figure 6-1b).

Electric fields are easily reduced, or shielded, by conductive objects. Intervening trees and buildings can cause substantial shielding. Magnetic fields, on the other hand, are not easily shielded except by special materials. As a result, the magnetic fields produced by charges moving in power lines or even wires in the home are not reduced when they penetrate most objects, including buildings and people.

Field Sources and Levels

Electric and magnetic fields are produced by the generation, transmission, and use of electricity. Anything in the path of this power transmission is a potential exposure source, from the generator, to the power lines, to an electric drill or clock. The home environment generally has magnetic field levels of the order of 0.1 µT (micro-Tesla). However, people can be exposed to fields as great as those near power lines when they use common appliances, such as hair dryers and electric shavers, which may produce magnetic fields of about 10 µT in their very immediate vicinity (Table 6-1). For reasons that need not concern us here, the strength of fields declines approximately linearly with distance as one moves away from power lines, with distance squared away from a single wire, and with distance cubed away from most small appliances. Some occupational environments, such as those using electric induction heaters, may have fields of 10-100 milli-Teslas.

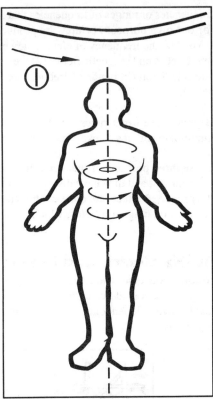

Figure 6-1b: Current flow in a person due to the magnetic field under an electric power line.

Table 6-1: Magnetic field levels (µT) at 60 Hz near various appliances

Appliance	3 cm	30 cm	1 m
Can Opener	1000 - 2000	3.5 - 30	0.07 - 1
Hair Dryer	6 - 2000	0.01 - 7	<0.01 - 0.3
Electric Razor	15 - 1500	0.08 - 9	<0.01 - 0.3
Mixer	60 - 700	0.6 - 10	0.02 - 0.25
Television	2.5 - 50	0.04 - 2	0.01 - 0.15
Coffee Maker	1.8 - 25	0.08 - 0.15	<0.01
Refrigerator	0.5 - 1.7	0.01 - 0.25	<0.01
Electric Drill	400 - 800	2 - 3.5	0.08 - 0.2

The strongest electric fields—near transmission lines—are typically about 10 kV/m (kilovolts/meter) directly under the conducting wires and 1 or 2 kV/m at the edge of the transmission line right-of-way. By contrast, 50-60 Hz electric fields from building wiring, power tools, and other electrical appliances typically range only up to 100 V/m.

Occupational sources and source characteristics are as varied as the work environments in which they are found, but two basic facts may be noted: sources of strong electric fields will be associated with the presence of electrical charge, such as around high-voltage equipment, and sources of high magnetic fields will generally be characterized by high currents, such as around high-amperage equipment or locations of high current flow. These need not be associated with heavy industry: electrical transformers and wiring located in vaults in office buildings are non-industrial sources of high fields.

Interaction With Humans

Power-frequency electric fields interact with humans through the outer surface of the body, inducing fields and currents within the body. Electric fields greater than 10 kV/m may cause hair vibration, which can be felt on sensitive areas such as the face, arms, or legs. Fields may also induce currents in metal structures, such as fences or automobiles, and these can produce shocks when humans contact the structure and provide a path to ground.

An electric field will cause currents to flow in the body, as expressed by a form of Ohm's law, $\mathbf{J} = \sigma \mathbf{E}$, where J is the induced current density in A/m^2 (Amperes/meter^2), σ is the tissue conductivity in S/m (Siemens/meter) which is about 0.2 for humans, and E is the electric field strength in V/m. A grounded person in an electric field experiences a short-circuit current of approximately:

$$I_{sc} = 15 \times 10^{-8} f W^{2/3} E_o$$

where I_{sc} is the short-circuit current in μA, f is the frequency in Hz, W is the weight in grams, and E_o is the external electric field strength in V/m. Thus, a person weighing 70 kg would have a total short-circuit current of about 153 μA in a 10 kV/m field. Researchers have investigated current flow in models of humans and laboratory animals exposed to 60-Hz electric fields. Their data indicate that current densities induced in a grounded, erect person exposed to a 10 kV/m vertical electric field are 0.55 μA/cm^2 through the neck and 2 μA/cm^2 through the ankles.

Time-varying magnetic fields can produce currents in loops in direct proportion to the magnetic field strength, the frequency of oscillation, and the radius of the current loop. A vertically directed field will cause current to flow through standing humans in loops whose plane is perpendicular to the vertical axis, or parallel to the ground. The current density can be expressed as $J = \sigma \pi f r B$, where f is the frequency and r is the radius of the loop. With average tissue conductivity equal to 0.2 S/m, the current density at the perimeter of the torso of an adult is approximately:

$$J = 0.1 \, f B$$

where J is the current density in A/m^2, f is the frequency in Hz, and B is the amplitude of the magnetic flux density in T (Teslas). The maximum current density induced in the normal residential environment is of the order of $\mu A/m^2$. For people using electric arc welders it may be one thousand times higher, of the order of mA/m^2.

Health Effects of Power-Frequency Fields

Early studies of possible health effects from exposure to power-frequency electric and magnetic fields were published in the former U.S.S.R. in the mid-1960s. Workers in electric switch yards reported a variety of subjective complaints, including problems with their cardiovascular, digestive, and central nervous systems. A study of electric utility linemen in the U.S. at about the same time failed to find any adverse health effects. Similar studies of electric utility workers continued for the next decade, with similarly contradictory results. In the early 1980s, however, attention broadened to workers whose jobs fit the category, "electrical occupation," and a number of studies suggested an increased risk, primarily of leukemia, for "electrical workers" in such varied occupations as electrical engineer, electronics engineer, television and radio repairman, electrician, motion picture projectionist, and telephone lineman. The actual exposures of these workers to electric and magnetic fields were not known, so it was impossible to link exposures to fields directly with health effects. However, the studies were suggestive enough to spark a great deal of research in this area.

To be able to evaluate health effects fully, the observational, or epidemiologic studies such as those described above must be accompanied by controlled laboratory experiments. Two types of laboratory experiments are important. *In vitro* (literally "in glass") studies using cells or tissue yield insight into the actual mechanisms that cause the observed effects. *In vivo* studies using whole animals confirm the possible occurrence of such effects, and help construct a dose-response relationship.

Effects in Cells and Tissue: *In Vitro* Studies

One *in vitro* study that fostered a great deal of follow-up laboratory and theoretical work investigated the influence of electromagnetic waves on ions in brain cells. In particular, chick brains were bisected and then bathed in a solution that contained a radioactive isotope of calcium, $^{45}Ca^{2+}$. The brain tissue incorporated some of the $^{45}Ca^{2+}$, which could then be used to trace the movement of calcium ions to or from the brain tissue. Next, one half of a brain was exposed to a 147-MHz electromagnetic wave that was amplitude modulated at 16 Hz; the other half was not exposed and served as a control. Both halves were in a solution that contained nonradioactive calcium. Compared to the unexposed half, the exposed half had about a 20 percent increase in calcium exchange with the physiologic solution in which it was immersed. When the researchers tried exposing the brains to a 16-Hz electric field, without the 147-MHz carrier frequency, they observed the opposite effect, namely, the exchange of calcium from the exposed half was decreased relative to the unexposed half.

Other researchers have tried to repeat, or replicate, the experiments and have obtained opposite results. Some researchers have replicated the experiments and found no effects. Some studies have suggested that the effect depends on the relative orientation and strength of the Earth's static magnetic field, and that it occurs in bands or "windows" of field frequency and intensity. This last finding is contrary to the commonly accepted idea that the greater the exposure the greater the effect; for example, higher levels of air pollution lead to more respiratory disease.

Attempts to explain the laboratory observations have focused on a resonance phenomenon known as cyclotron resonance. Experiments were designed and conducted to test cyclotron resonance as an explanation, and some of the experimental results appear to confirm it. However, theoretical problems remain. First, the calcium ions are hydrated, making their effective mass much greater than that which fits the requirements for cyclotron resonance. Second, the radius of the ion's orbit would be of the order of meters, rather than the size of cells.

Effects in Laboratory Animals: *In Vivo* Studies

Studies with laboratory animals have investigated the possible effect of electric and magnetic fields on a broad range of systems and outcomes, including body weight; hematology and immunology; the endocrine, cardiovascular, and nervous systems; circadian rhythm; behavior; genetics; reproduction and development; and cancer. In general, the results have been negative, inconsistent, or of limited relevance to questions of adverse human health effects. Studies that have received more attention because of their implications for gross human health effects are summarized below.

In one study, chicken embryos were exposed to 60-Hz sinusoidal electric fields of up to 100 kV/m. Chicken embryos were used because they are a rapidly developing, and hence, a relatively susceptible organism. Because they are stationary, it is easy to monitor the exposure levels. No effects were found on mortality, deformity, birth weight, or development.

A very large, follow-up study of chicken embryos in six different laboratories in North America and Europe points out the difficulty with conducting and interpreting the research on health effects of power-frequency fields. Two laboratories found a significant increase in abnormal embryos in the exposed groups; however, four did not. On the other hand, when the results from all six laboratories were analyzed together, the results were statistically significant.

Other *in vivo* studies have considered the effects of power-frequency fields on mammals. Exposing pregnant rats to electric fields of 10, 65, and 130 kV/m produced no significant increases in litters with malformations compared to pregnant rats that were not exposed. Similarly negative results were obtained for rats exposed to 60-Hz magnetic fields of 0.61 and 1000 μT. Finally, exposing mice that had leukemia to fields of 500 μT did not affect how long they survived.

Effects in Human Populations: Epidemiological Studies

It is not possible to do experiments with human beings in the same way that can be done with animals. Thus, well-controlled experiments on health effects, in which only one parameter, such as exposure to electric and magnetic fields, is allowed to vary, are not normally available as research tools. Information about the health effects of agents in the environment can be obtained, however, using a form of inquiry known as epidemiology. Literally the study of epidemics, the epidemiologic method is to look at large groups of people. By attempting to group people according to their difference in a single aspect, such as their exposure to electric and magnetic fields, other uncontrollable differences between them may "average out" because of the large numbers. If there is a link between exposure and a health effect, then the population that is more highly exposed would presumably have a greater proportion of people with the effect.

The results of some epidemiologic studies of the health effects of electric and magnetic fields are described below. More than any other method of inquiry, it is the epidemiologic studies that continue interest in possible health effects of low-frequency field exposures.

A 1979 study in Denver, Colorado, suggested an association between the type and proximity of electric utility wiring outside the home and the risk of childhood cancer. The risk increased with closer and thicker wires, which presumably carried higher currents, and which would lead to higher magnetic fields at the residence. A later study reported an increased risk for all cancers in adults who were similarly exposed.

Many epidemiologic studies have followed that initial report, and it would be difficult to overstate the research interest that it fostered. Among the many subsequent studies, two were conducted essentially as follow-up work. One was to replicate the childhood cancer study. It included actual measurements of magnetic fields in addition to basing exposures on the type of wire near the home. Although there again was a significant association between wiring type and cancer, there was not a significant association with the actual measured values of field strength. The second study, conducted in Seattle, looked at adult leukemia, and it did not demonstrate a significant association with either wire types or measured field values.

Another study of childhood leukemia, carried out in Los Angeles, also used both wire codes and measurements to assess exposures to magnetic fields. It, too, found a statistically significant risk associated with wire codes but not with measured fields.

The recurring, statistically significant association of cancer with wire types but not with measured fields suggests one of the following may be true: (1) the fields do cause adverse health effects, and basing exposure on wire types is a better indicator of what the true cause is than are the measurements; (2) the fields do not cause adverse health effects, but basing exposure on wire types better indicates the true causal environmental factor, for example, air pollution due to local traffic density; or (3) the study results are essentially "noise." As studies continue to find a significant association with surrogates for exposure but not with measured fields, the third possibility becomes less plausible. Continued research in this area with exposure assessments based fully on measurements will be necessary to resolve the apparent contradictions.

Exposure Guidelines

Ideally, exposure guidelines and standards are established on the basis of an accepted mechanism of interaction, dose response studies in animals, and epidemiological evidence of similar effects in humans. None of this has occurred for 60-Hz fields. However, because of concerns from workers and the general public, exposure guidance has been developed by a number of countries and organizations. The American Conference of Governmental Industrial Hygienists (ACGIH) in the U.S., the National Radiological Protection Board (NRPB) in the U.K., the German government, and the International Non-Ionizing Radiation Committee of the International Radiation Protection Association (IRPA/INIRC) have all developed exposure standards or guidelines. With contradictory or nonexistent guidance from health effects studies, how is this done?

The exposure limit rationale is based on induced currents in the body and on an accepted biological effect (not an adverse health effect) in humans. The effect, which is reproducible and well-documented, is the production of magnetophosphenes, or the sensation of flickering light in the eye. First described by d'Arsonval in 1896, the effect has been investigated extensively. The minimum magnetic field that will produce phosphenes is approximately 10 mT at a frequency of 20 Hz. The required electric field is about 40 kV/m at that same frequency. Other frequencies require stronger fields. The threshold current density is estimated to be 20 mA/m^2. Because phosphenes are the physiologic effect that can be reliably replicated at the lowest field levels, exposure guidance, for fields *per se*, has been based on avoiding their production in exposed humans by maintaining induced current densities below 10 mA/m^2. Both the ACGIH and the INIRC developed guidelines by limiting induced current densities in the body to those levels that occur normally, i.e. up to about 10 mA/m^2 (higher current densities can also occur naturally in the heart). They acknowledged that biological effects have been demonstrated in laboratory studies at field strengths below those permitted by the exposure guidelines, but both agencies concluded that there is now no convincing evidence that exposure to these field levels leads to adverse health effects.

Radio-Frequency and Microwave Radiation

Another portion of the electromagnetic spectrum that has been captured for human purposes is that covered by radio-frequency (RF) and microwave (MW) radiation. The frequency ranges from about 10 kHz to 300 GHz. A wide variety of sources, including AM and FM radio, television, cellular telephones, radar, walkie-talkies, and microwave ovens produce and use electromagnetic waves within this broad range of frequencies.

The primary effect of these waves when they interact with people is heat. This arises from interactions with polar molecules, such as water, as energy is imparted from the waves to the molecules. Consideration of the effect of microwave ovens, most of which operate at a frequency of 2450 MHz, should make this clear.

Guidance on safe levels of RF and MW exposure is based on the rate that energy is transferred to the body. Normal metabolism, the act of living, produces about 1 joule/second/kilogram or 1 W/kg in the body. For the general population, not exposed at work, the National Commission on Radiation Protection and Measurements (NCRP) has recommended that the energy absorbed from exposure to RF or MW radiation not exceed a rate of 0.08 W/kg, averaged over 30 minutes, which is well within normal metabolic levels.

However, people, and biological tissue, do not respond with the same efficiency to all frequencies, and as a result, guidance on safe levels of RF and MW exposure differs with frequency. The transfer of energy to the body is most efficient when the length of the person is about 40% of the wavelength of the radiation.

Ultraviolet Radiation

Although the major source of ultraviolet, or UV, radiation is the sun, human activities also produce ultraviolet radiation. For example, UV is used for germicidal and "black light" lamps, and in tanning booths. It interacts with people differently from the previously discussed classes of electromagnetic waves and produces the well-known effect of skin erythema, commonly known as sunburn.

The previously mentioned power-frequency fields, and radio-frequency and microwave radiation, are referred to by their frequency. On the other hand, ultraviolet (as well as visible and infrared) radiation is discussed in terms of its wavelength. The spectrum of ultraviolet is divided into three regions, as shown in the following chart.

Region	Wavelengths	Description
UV-A	315 - 400 nm	Blacklight
UV-B	280 - 315 nm	Sunburn
UV-C	100 - 280 nm	Germicidal

Ultraviolet light penetrates the skin to different degrees, depending on wavelength. Wavelengths shorter than about 295 nm (nanometers) are absorbed almost completely in the outer layer of skin, or the epidermis. Longer wavelengths penetrate more deeply. In the eye, the short-wavelength UV-C and UV-B are absorbed in the cornea; longer wavelengths penetrate to the lens. At normal levels, UV does not cause heating of body tissues. Instead, photochemical effects occur when the UV energy is absorbed by molecules, including DNA, in the body.

A typical response in the eyes is photokeratitis, commonly known as snow blindness or welder's flash, depending on the source. The symptoms include a feeling of sand in the eyes, and they typically do not occur until 6 to 12 hours after exposure. A much more serious effect is the production of cataracts in the lenses of the eyes. The peak response occurs at about 300 nm, and experiments have produced cataracts in animals after both acute exposures (high intensities of UV for a short time) and chronic exposures (low intensities for a long time), suggesting that the use of sunglasses that block UV may offer some effective long-term protection against cataract formation.

As must be apparent from the discussion so far, different wavelengths (or frequencies) of non-ionizing radiation interact differently with the body to produce different effects: extremely low frequency fields can cause magnetophosphenes, radio-frequency and microwave radiation can cause heating, and UV can lead to sunburn. In addition to such differences, which occur across wavelength changes of many orders of magnitude, the body's response can also vary greatly over a relatively narrow band of wavelengths, as described above, for sunburn. The relative response of the body (or the relative effectiveness of the radiation) is referred to as the action spectrum. Guidance on safe exposure to UV light includes consideration of the action spectrum for erythema (Figure 6-2).

Sunburn is the most familiar effect of overexposure of the skin to UV. It occurs in response to wavelengths from 290 to 320 nm, while the most effective wavelengths are those near 297 nm. The minimum UV exposure that will produce sunburn is called the minimum erythema dose, or MED. At 297 nm it is a radiant exposure of about 140 J/m^2 for fair-skinned individuals.

Longer-term effects on the skin include premature aging and skin cancers. Premature aging is characterized by a leathery appearance of the skin. Skin cancers can be divided into two types: nonmelanoma and melanoma. Nonmelanoma skin cancers are produced primarily by UV-B exposure. Development of nonmelanoma skin cancers is more likely among fair-skinned individuals living at lower latitudes and higher elevations, i.e. where UV intensities are higher. Melanoma is a cancer of the cells in the skin that produce pigment and is more likely to develop on the trunk in men and the lower legs in women. Thus, it is not so directly related to high UV exposure as are the nonmelanoma forms. It is more common among people with blue, gray, or green eyes, and in urban rather than rural locations.

Certain chemicals can increase one's sensitivity to UV from the sun. The antibiotic tetracycline is one, and those taking it are advised to avoid exposure to the sun because they will burn more easily. Lime juice and some deodorant and antibacterial agents in soaps will also cause photosensitivity.

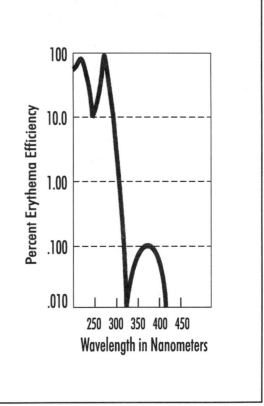

Figure 6-2: **Erythema Action Spectrum for Human Epidermis**

SOCIETY'S REACTIONS TO THESE ENVIRONMENTAL ISSUES

The most prevalent source of electromagnetic fields and energy in society are high voltage lines for the transmission of electricity. Like many environmental threats facing the United States today, the threat posed by electromagnetic radiation is a by-product of a process, the generation of electricity, whose benefits to society far outweigh the risks associated with it. The transmission of electricity through high voltage lines has always been accompanied by some degree of controversy, since the negative impacts of high power lines have always been known: danger of contact, noise, interference with electronic signals, impairment of beauty and privacy, and limitations on the value and sale of adjacent land. In recent years, however, the possible health threats presented by electromagnetic fields has focused more attention on high voltage lines and affected the process by which society regulates such lines and, consequently, the electric power industry.

History of Government Regulation of High Voltage Lines

The provision of electrical power by public utilities is a highly regulated industry, with regulation being accomplished almost exclusively at the state government level. Individual states may therefore approach, and regulate, public utilities in different ways, although the general scope of activities that are regulated remains the same throughout the country. Regulations control, among other issues, the geographical area in which a utility can operate, the customer base it must serve, the rates it may charge, and the rate of return its investors may earn.

In addition to the state agencies (generally known as Public Utility Commissions, or PUCs) which administer the regulations, utilities are also subject to supervision by state courts, both directly and through judicial review of agency determinations.

In addition to the existence of a clear causal connection between the presence of high voltage lines and certain potential dangers, the possibility that biological hazards could be caused by the electromagnetic fields associated with high voltage power lines has been addressed as described earlier in this chapter. Some states have reacted by ordering further study and have required utilities to survey electromagnetic field strengths generated by existing facilities and/or newly constructed facilities. Most states have increased the type and amount of regulations applicable to such power lines. In some instances, these new regulations restrict the strength of electric or magnetic fields which new facilities may generate; other states have affirmatively stated that no such restrictions are necessary.[1] No state has required modification of existing facilities, nor has any state required the restriction of field strengths generated by any facilities other than power lines, such as distribution facilities. Although further regulatory limits on high voltage lines (and the associated electromagnetic fields) have yet to be enacted through the legal process, those who are potentially most affected by the placement of high voltage lines (that is, proximate property owners) can attempt to influence the location of the lines or persuade the industry to find ways to reduce the emission of electromagnetic fields. One opportunity to make such an impact is presented when a utility is attempting to locate new high voltage lines.

Society's Ability to Affect the Placement of High Voltage Lines

In particular, the siting of high voltage lines requires the successful completion of a series of complicated requirements and satisfaction of legal and technical criteria. Approval for the construction of new electrical facilities requires that the utility demonstrate, for example, that the new facility is needed for the efficient and economic delivery of adequate power to its service area, and that the utility has properly taken into account the environmental impact of the project. These considerations may also be necessary in order to satisfy the requirements of the eminent domain statutes which give public utilities the right to seize privately held property to obtain ownership of the land on which the electrical facilities will be constructed. After this initial hurdle and the selection of the sites for the power line route, the PUC then notifies affected property owners and holds public hearings at which the utility explains the need for the facilities and the PUC receives the input of the public on such issues as the existence of alternative routes for the power lines. This represents the first substantial opportunity for those with the potential to be most affected by the electromagnetic fields generated by high voltage lines to make their concerns known and present arguments and comments, and any documentation relevant to the issue.

After consideration of public opinion, the PUC has the authority to issue final approval of the project and may do so, despite the objection of local residents, provided that the PUC is not acting in an arbitrary and capricious manner in issuing the approval. The only activity which a PUC cannot undertake with respect to a utility is the actual operation of the utility, although the PUC may be given the power to so act in the event that the utility is determined to have abused its discretion.

Once the PUC issues final approval regarding the siting of the lines, the property on which the high voltage lines will be located must be acquired from the persons or entities who currently own it. The process by which any government or division thereof obtains title to privately held property is known as condemnation. An understanding of the condemnation process is necessary, as it provides another significant forum in which those affected by the placement of high voltage lines can influence, although not prevent, the placement.

Condemnation Process

The process known as "condemnation" grows out of the government's right, known as "eminent domain," to seize or take land for the use of the public. The eminent domain right is considered to be an "inherent attribute of sovereignty,"[2] having its origins in ancient Roman times. This right of government is long beyond challenge; in fact, the Fifth Amendment to the U.S. Constitution contains only a confirmation of the existence of the right rather than an actual grant of the right.[3] In the United States, the government's ability to seize privately owned land is, however, somewhat limited by the Fifth Amendment's requirement that the government provide the property owner with just compensation for the property seized. The actual legal process by which the seizure occurs and title is transferred to the government constitutes the process of condemnation.

Property being condemned can represent a separate parcel of property, or it can represent only a portion of a piece of property. The property owner who retains the remaining portion of the non-condemned piece of property, or the owner of the adjacent but separate piece of property, may still be affected by activities on the condemned property even after he no longer owns it. The condemnation process allows property owners to reject the government's offer of compensation and oppose the propriety of the condemnation process by alleging that the government failed to satisfy all requirements imposed by condemnation regulations and/or the regulations governing the placement of high power lines, and by claiming that the high-voltage facilities will result in detrimental impact to the community or neighboring property. Before the issue of the biological risks of electromagnetic fields was raised, the primary concerns raised by proximate property owners were those previously mentioned: danger of contact, noise, interference with electronic signals, impairment of beauty and privacy, and limitations on the value and sale of adjacent land.

These detrimental impacts are generally not of a sufficient magnitude to warrant relocation of high-voltage lines. Rather, these issues, if validated, may be incorporated into the calculation of compensation to the property owner or may result in certain accommodations to the community (for example, specifications on the size and type of fence surrounding the high-voltage facilities, requirement to provide aesthetic plantings or buffers, etc.). It can be said that the benefit to society of having high voltage lines (that is, society's need for power) outweighs the individual's rights to have the voltage lines placed to his liking, although he may be entitled to some financial compensation for the impact associated therewith. Courts which must review the determination of a condemning authority as to the property to be condemned offer the authority wide discretion and will generally not challenge the authority's choice of property. If the PUC has not acted arbitrarily or irrationally in its siting determination, the determination will stand. While courts differ on what type of impacts the PUC must have considered in order for the PUC to have acted in a rational manner, there is no standard for what constitutes adequate consideration; if the PUC "considered" the issues in its decision-making process, the PUC has not acted in an arbitrary manner.

How Biological Risk Posed by EMF Affects the Legal Process

With the epidemiologic studies as a foundation, there has, in recent years, been claims that electromagnetic fields cause biological effects and health hazards. Although high voltage lines are by no means the only source of electromagnetic fields, they represent the most obvious target. Despite the lack of government action (that is, through additional regulation) in responding to the research which suggests, but does not unambiguously prove, a causal connection between electromagnetic fields and certain diseases,[4] litigation based on the existence of disease or injury caused by exposure to electromagnetic fields may nevertheless be successful. Such claims must, however, be based on actual injury and not just the fear of future injury.[5] Obviously, claims based on the occurrence of an actual disease or injury arise after high-voltage lines are constructed; the focus of litigation becomes the utility company who owns, operates, and maintains the power lines. Such litigation has not become common, primarily due to the difficulty in proving causation. Not only is the scientific data insufficient to establish a causal connection between exposure to electromagnetic fields produced by high voltage lines and disease or injury, but the claimant must also show the absence of other possible sources, including other electromagnetic fields and other environmental conditions.

The scientific data presents a stronger basis when used to support a property damage claim. The actual existence of health hazards is not necessary in order for a property owner to prevail in a suit against the utility for property damage. Instead, the property owner must demonstrate that a significant segment of the buying public either believes in the existence of such hazards or there is significant uncertainty that people will not pay as much for property located next to power lines as for a comparable property not located next to power lines.[6] In these cases, it can be said that the property owner is demonstrating that there is a "stigma" associated with the property and that the property value has therefore been adversely affected. The property owner can be compensated by the utility for the damages represented by this reduction in property value. These damages, however, hardly equate to a resolution of the concern over the health effects of electromagnetic fields posed to those living near the power lines.

RELEVANT EDUCATIONAL AUDIO-VISUAL MATERIALS

Note: In some cases, A/V materials can be previewed before purchasing or renting.

Electromagnetic Waves (1987) 21-minute videotape. The vast array of electromagnetic radiation that surrounds us is explored. Available from: Barr Films, 12801 Schabarum, Irwindale, CA 91706.

Educational materials on 60-Hz fields, including a slide presentation, are available through the "Resource Paper" series from the Electric Power Research Institute, P.O. Box 10412, Palo Alto, CA 94303.

Local electric utilities will also have materials and possibly speakers.

INTERNET RESOURCES

Note: URLs are subject to change. Access the following sites on the World Wide Web for further information.

Electric and Magnetic Fields (EMF) Research and Public Information Dissemination (RAPID) Program
http://www.niehs.nih.gov/emfrapid/home.htm

The National Institute of Environmental Health Sciences (NIEHS) and the Department of Energy (DOE) are coordinating the implementation of the Electric and Magnetic Fields (EMF) Research and Public Information Dissemination (RAPID) Program, established by the 1992 Energy Policy Act (Section 2118 for Public Law 102-486) which was signed in October 1992. This is a five-year United States federally coordinated effort to evaluate developing technologies and research on the effects on biological systems of exposure to 60 Hz electric and magnetic fields produced by the generation, transmission and use of electric energy and to communicate these results to the public sector.

This site is an excellent resource for information about current research projects in EMF that are being conducted around the nation. It also includes an EMF Measurements Database, a "Questions and Answers" section about EMF in the workplace, information about relevant conferences, and other resources.

EMF Health & Safety Digest
http://www.rsba.com/hsd/hsd_highlight_f.shtml

Since 1983, *EMF Health & Safety Digest* has provided readers with up-to-date, in-depth and reliable coverage of EMF issue developments ranging from research to governmental and judicial actions to local controversies. This site provides highlights of the most recent issue of the *Digest*, as well as ordering information.

EMF-Link ®
http://infoventures.microserve.com/emf/

EMF-Link ® provides substantive information on biological and health effects of electric and magnetic fields (EMFs) from common sources such as power lines, electrical wiring, appliances, medical equipment, communications facilities, cellular phones, and computers. Full access to EMF-Link ® is obtained by subscribing to one of Information Ventures, Inc.'s publications and registering online for access. However, a great deal of information, including FAQs about EMF topics, is available through the section users may access at no charge.

The World Health Organization's International EMF Project
http://www.who.ch/programmes/peh/emf/emf_home.htm

In collaboration with international agencies and organizations, WHO is pooling resources and knowledge concerning effects of exposure to EMF and making a concerted effort to 1) identify gaps in knowledge, 2) recommend focused research programs that allow better health risk assessments to be made, 3) conduct updated critical reviews of the scientific literature, and 4) work towards an international consensus and resolution on the health concerns. The EMF project was established by WHO in 1996.

This site is an excellent resource for an international perspective to EMFs. It provides details of the organization's current and long-term activities; results of scientific meetings, current and future events, and publications; a fact sheet on EMF; and links to related sites. Linked sites include the International Commission on Non-Ionizing Radiation Protection, the Swedish Guidance Document, the International Labour Office, and more. In addition, the WHO site lists other organizations or societies with an interest in EMF exposure and health but not associated with the Project. These include: Microwave News; Bioelectromagnetics Society; and the Electromagnetic Energy Association.

BIBLIOGRAPHY

ACGIH. 1990a. Notice of intended change—sub-radio-frequency (30 kHz and below) and static electric fields. *Applied Occupational and Environmental Hygiene* 5:734-737.

AIHA Publications. "Extremely Low Frequency (ELF) Electric and Magnetic Fields." American Industrial Hygiene Association, 2700 Prosperity Avenue, Suite 250, Fairfax, VA 22031, 1995. ISBN 0-932627-61-7.

AIP. 1997. "Research Council Panel Tries to End Controversy Linking EMFs with Cancer and Other Health Disorders," *Physics Today*, Jan. 1997, p. 49.

ACGIH. 1990b. Notice of intended change—sub-radio-frequency (1 Hz to 30 kHz) magnetic fields. *Applied Occupational and Environmental Hygiene* 5:884-892.

AIBS. 1985. Biological and Human Health Effects of Extremely Low Frequency Electromagnetic Fields: Post 1977 Literature Review. Technical Report AD/A152 731, prepared for the Naval Electronic Systems Command. American Institute of Biological Sciences, Arlington, VA.

Bennett, W.R., Jr. "Cancer and Power Lines," *Physics Today*, April 1994, p. 23.

Bennett, W.R., Jr. *Health and Low-Frequency Electromagnetic Fields*. Yale University Press, New Haven, 1994.

Department of Engineering and Public Policy. "Electric and Magnetic Fields from 60 Hertz Electric Power: What do we know about possible health risks?" Carnegie Mellon University, Pittsburgh, PA 15213, 1989.

Department of Engineering and Public Policy. "Fields from Electric Power." Carnegie Mellon University, Pittsburgh, PA 15213, 1995.

"Electromagnetic Fields," *Consumer Reports*, May 1994.

Hitchcock, R.T., R.M. Patterson. *Radio Frequency and ELF Electromagnetic Energies: A Handbook for Health Professionals*. Van Nostrand Reinhold, New York, 1994.

IEEE. 1987. IEEE standard procedures for measurement of power frequency electric and magnetic fields from AC power lines. IEEE Std 644-1987. The Institute of Electrical and Electronics Engineers, Inc., New York.

IRPA/INIRC. 1990. Interim guidelines on limits of exposure to 50/60 Hz electric and magnetic fields. International Non-ionizing Radiation Committee of the International Radiation Protection Association. *Health Physics* 58:113-122.

Murray, W.M., R.T. Hitchcock, R.M. Patterson, S.V. Michaelson. "Nonionizing Electromagnetic Energies." In *Patty's Industrial Hygiene and Toxicology: Volume 3, Part B, Third Edition*, ed. L. Cralley et al. John Wiley & Sons, New York, 1994.

NRPB. 1986. Advice on the protection of workers and members of the public from the possible hazards of electric and magnetic fields with frequencies below 300 GHz: a consultative document. p. 2. National Radiological Protection Board, Chilton, Didcot, Oxfordshire, OX11 0RQ, U.K.

NRPB. 1991. Biological effects of exposure to non-ionizing electromagnetic fields and radiation: I. Static electric and magnetic fields. Report NRPB-R238, National Radiological Protection Board, Chilton, Didcot, Oxfordshire, OX11 0RQ, U.K., July 1991.

NRPB. 1991. Biological effects of exposure to non-ionizing electromagnetic fields and radiation: II. Extremely low frequency electric and magnetic fields. Report NRPB-R239, National Radiological Protection Board, Chilton, Didcot, Oxfordshire, OX11 0RQ, U.K., July 1991.

NRPB. 1991. Biological effects of exposure to non-ionizing electromagnetic fields and radiation: III. Radio-frequency and Microwave Radiation. Report NRPB-R240, National Radiological Protection Board, Chilton, Didcot, Oxfordshire, OX11 0RQ, U.K., December 1991.

NRPB. 1992. Electromagnetic fields and the risk of cancer. Documents of the NRPB, 3(1), National Radiological Protection Board, Chilton, Didcot, Oxfordshire, OX11 0RQ, U.K.

ORAU. 1992. Health effects of low-frequency electric and magnetic fields. Oak Ridge Associated Universities. Report GPO#029-000-00443-9, U.S. Government Printing Office, Washington, D.C., and ORAU 92/F-S, National Technical Information Service, Washington, D.C.

Patterson, R.M., R.T. Hitchcock. "Non-ionizing radiation and fields." In *Health and Safety Beyond the Workplace*, ed. L.V. Cralley et al. John Wiley & Sons, Inc., New York, 1990.

U. S. Department of Energy. "EMF in the Workplace." Report DOE/GO-10095-218, September, 1996.

WHO. 1984. Environmental Health Criteria 35: Extremely Low Frequency (ELF) Fields. World Health Organization, Geneva, Switzerland.

WHO. 1987. Environmental Health Criteria 69: Magnetic Fields. World Health Organization, Geneva, Switzerland.

DISCUSSION QUESTIONS

1. What is the wavelength of power-frequency electric and magnetic fields?

2. What would be your short-circuit current if you were standing in a 5 kV/m electric field?

3. What would be the current density at the perimeter of your torso if you were standing in a vertical, 60-Hz magnetic field of 3 µT?

4. What would be the current density at the perimeter of your head at the ACGIH guidance level for 60-Hz magnetic fields, 1 mT?

5. What is the frequency of non-ionizing radiation that interacts with you most efficiently (the resonant frequency for you)?

6. The energy in a photon of non-ionizing electromagnetic energy depends directly on the frequency. Compare the relative energy in a photon of electromagnetic radiation from a local radio station with that in the frequency of UV light that is the most effective in causing sunburn. If the energy of the radio station photon were worth a dollar, what would be the value of the UV photon?

7. Find the frequency of the microwave oven you use most frequently.

8. Research the occupational exposure limits for 60-Hz electric and magnetic fields recommended by the IRPA, the ACGIH, and the NRPB.

9. The designation "SPF" describes the effectiveness of sunscreen. What do the initials "SPF" mean? Is there a scientific basis for assigning SPFs?

10. Research the different chemicals used in sunscreen to absorb UV radiation before it interacts with the skin. What different chemicals are used? How does the UV radiation interact with the chemicals? Are some better than others? Overexposure to UV can cause sunburn and even skin cancer, and it is certainly hazardous. Are there possible hazards from using sunscreen?

NOTES

1. Young, Sherry, "Regulatory and Judicial Responses to the Possibility of Biological Hazards from Electromagnetic Fields Generated by Power Lines," *Villanova Law Review 36*, p. 129-190.

2. Dukeminier, Jesse and James E. Krier, *Property*, Little, Brown and Company, page 1094 (1981).

3. See *United States v. Carmack*, 329 U.S. 230 (1946).

4. See, for example, Office of Technology and Assessment, U.S. Congress, "Biological Effects of Power Frequency Electric and Magnetic Fields—Background Paper," OTA-BP-E-53 (May, 1989); and "New York State Power Lines Project Scientific Advisory Panel, Biological Effects of Power Line Fields" (July, 1987), Final Report prepared for the New York State Power Lines Project, Albany, NY.

5. See, for example, *Pappas v. Alabama Power Co.*, 270 Ala. 472 (1960), *Central Ill. Light Co. v. Nierstheimer*, 26 Ill.2d 136 (1962).

6. See, for example, *Louisiana Power & Light Co. v. Mobley*, 482 So2d 706 (La. Ct. App. 1985), *San Diego Gas & Electric Co. v. Daley*, 205 Cal.App.3d 1334 (Cal. Dist. Ct. App. 1988).

7
The Portrayal of Environmental Issues in Film and in Non-Documentary Television Programs

Much of the public's knowledge of science, scientists, and technology comes from the mass media—primarily newspapers, magazines, television, and film. Television and film are perhaps more influential than other media in creating public perceptions of these subjects because of their visual imagery. If we hear the word "scientist," for example, an image from visual media usually comes to mind, rather than a more abstract conception of a real-life scientist about whom we've read or heard.

Therefore, it is imperative to study how the visual media depict science, scientists, and technology, in order to better understand how such portrayals may be influencing our perceptions. For example, in the media, scientists are commonly portrayed as being so far removed from everyday society that their work has no meaning for us. Even more familiar is the "mad" scientist, usually male, tinkering in his lab on experiments that pose a threat to humanity.

The films and television programs listed in this chapter do not include examples of the above-mentioned stereotypes. They do, however, include examples of another type: applied science and technology (or the mad scientists' experiments) gone wrong. Each film or program includes a depiction of an environmental issue detailed in the preceding chapters: nuclear power, indoor air pollution, the greenhouse effect and stratospheric ozone depletion, or electromagnetic waves. In many instances, the issue illustrated is a problem resulting from interference with the natural "way of the world"—in methods as serious as nuclear energy experiments to those as seemingly innocuous as preservatives and other chemicals in product packaging.

After viewing the film relevant to your study, take a few minutes to reflect on the messages, both obvious and subtle, that the film conveys about science and technology. It is also interesting to note the similarities and differences between various films' portrayals of a particular topic or of science in general.

The availability on videotape of the films listed below will vary from time to time. The *Star Trek: The Next Generation* episodes are being released periodically.

The scientific accuracy of these films or television programs varies considerably. Assessing the accuracy or inaccuracy of the science depicted in one or more of these films or television programs could serve as an interesting research project. As an example, the film *Them!* illustrates a theme of many science fiction films (particularly those of the 1950s)—that radiation fall-out may create giant monsters. This is scientific nonsense! One of the many reasons that giant ants can not exist is that insects breathe through holes in their sides: oxygen molecules diffuse through these holes. Since the diffusion time of oxygen molecules is proportional to the square of the distance that the molecules must travel, this diffusion time for a 100-inch-long ant would be $(100)^2 = 10,000$ times longer than that of a one-inch ant. In short, giant ants would suffocate.

For further discussions of the accuracy and inaccuracy of the science in science fiction films and popular television programs, see *Science in Cinema: Teaching Science Fact Through Science Fiction Films* by Leroy W. Dubeck, Suzanne E. Moshier, and Judith E. Boss (Teachers College Press of Columbia University, 1988), and *Fantastic Voyages: Learning Science Through Science Fiction Films* by Leroy W. Dubeck, Suzanne E. Moshier, and Judith E. Boss (American Institute of Physics Press, 1993).

FILMS RELATED TO CHAPTER 3: NUCLEAR POWER PLANTS, NUCLEAR WASTE DISPOSAL, AND OTHER RELATED ISSUES

The China Syndrome
(United States) 1979, color, 123 minutes
Credits: Director, James Bridges; Producer, Michael Douglas.
Cast: Jane Fonda (Kimberly Wells), Jack Lemmon (Jack Godell), and Michael Douglas (Richard Adams).

Plot Summary:
The China Syndrome is a thriller that raises disquieting questions about the safety of nuclear power plants. The events leading up to the accident in the film depict actual occurrences at nuclear plants, including the basic cause of the accident at Three Mile Island (which occurred after the film was released).

The key character in the film is Jack Godell, who is a shift supervisor at a major nuclear power plant in southern California. He supports the use of nuclear power, but when a turbine trip shakes his plant he becomes convinced that he felt another shock caused not by the turbine shutting down, but by something else rumbling within the plant. The shaking causes a near-accident because a needle becomes stuck on a roll of graph paper, and the operators think that they need to lower the level of the water over the nuclear pile. Actually, the level is already dangerously low. The film asserts that if the pile were ever uncovered, the result would be the "China Syndrome," so named because the heated nuclear materials would melt through the floor of the reactor plant and, in theory, keep on going until they hit China. In practice, there might be an explosion with a release of enormous amounts of radioactive material.

The accident takes place while a TV news team headed by a reporter, Kimberly Wells, is filming at the plant. She tries to get the story of the near-accident on the air but the plant's public relations people prevent it. Consequently, she and her cameraman, Richard Adams, investigate further. Meanwhile, Godell has been conducting his own investigation and learns that X-rays used to check key welds at the plant have been falsified.

The movie then takes off in classic thriller style. Godell desperately races against time to force the plant's management to check the welds before a catastrophe occurs.

The running time at which the reporter visits the nuclear power plant (during which the first near accident occurs) commences at 6 minutes after the film opens.

Silkwood

(United States) 1983, color, 128 minutes
Credits: Director, Mike Nichols; Producers, Mike Nichols and Michael Houseman.
Cast: Meryl Streep (Karen Silkwood), Kurt Russell (Drew Stephens), Cher (Dolly Pelliker), and Craig Nelson (Winston).

Plot Summary:
Silkwood is the story of workers in an Oklahoma nuclear plant who make plutonium fuel rods for nuclear reactors. It is based on the real-life story of Karen Silkwood. As the movie begins, Karen Silkwood seems to fit naturally into this assembly-line world. The nuclear plant in the film is behind on completing an important contract. Consequently, people are working overtime and some safety precautions are being ignored. A series of small incidents convinces Silkwood that these compromises are potentially dangerous, and that the health of the workers is being put at risk. For example, the company has a truck, contaminated by radiation, illegally dismantled and the pieces buried at night. Furthermore, the company is ignoring obvious falsification of safety and workmanship tests. When Silkwood is contaminated by radioactive waste products, she approaches her union. The union is facing a decertification challenge and thus sees some advantages in publicizing her complaints. She gets a free trip to Washington—her first airplane ride—and meets with union officials who seem more concerned with publicity than with her plant's working conditions.

Upon her return to the plant she becomes a committed union leader in the struggle to make working conditions safer. Her union wins its decertification election, but her continuing concern about radiation safety results in her becoming alienated from her co-workers. She is again contaminated with radiation; this time, she believes that someone deliberately contaminated her.

The real Karen Silkwood died in a mysterious automobile accident while driving to deliver some documents to a *New York Times* reporter. Was the accident actually murder? The movie doesn't suggest the answer nor does it point suspicion only toward Silkwood's employer, since there were many people angry at her.

The film depicts some safety violations 2 minutes after it begins; Silkwood learns about the truck being buried 20 minutes into the film; and she is contaminated with radioactive materials for the first time 38 minutes into the film.

Them!

(United States) 1954, black and white, 93 minutes
Credits: Director, Gordon Douglas; Producer, David Weisbart.
Cast: James Whitmore (Sgt. Ben Peterson), James Arness (Robert Graham), Joan Weldon (Dr. Patricia Medford), and Edmund Gwenn (Dr. Medford).

Plot Summary:

As the movie opens, a little girl is seen wandering in the desert. She is rescued by Sgt. Ben Peterson, who takes her to a nearby trailer only to find that one of its sides is pulled apart and sugar cubes are scattered about. The occupants are nowhere to be found. The little girl, who is in a trance-like state, is taken to a hospital while Peterson and another patrolman stop at a country store. There they find the owner dead and the place in shambles. Money is still in the cash register. The other patrolman is left on guard while Peterson goes to the hospital. The patrolman hears a strange, rapidly pulsating, high-pitched sound, goes outside the store to investigate, and we hear his scream and the firing of his gun.

The local authorities call in the FBI for assistance because of the disappearance of the patrolman. As Special Agent Robert Graham arrives, the coroner informs the police that the store owner not only has a broken back and skull, but his body contains enough formic acid to kill twenty men. The cast of an imprint found near the trailer is sent to FBI headquarters for identification.

Washington dispatches two scientists (Dr. Medford and his daughter, Patricia) who are experts on insects. At the hospital, they shock the little girl out of her trance by having her smell formic acid. The girl screams, "Them!"

The scientists then return to the trailer site where they encounter a ten-foot long ant, which is killed by a submachine gun. Dr. Medford speculates that the atomic bomb exploded ten years earlier, in 1945, at White Sands, New Mexico, may have caused mutations in the desert ants to produce these giants. This is the ominous scientific theme of the film—the unknown effects of nuclear radiation on other life forms.

The nest of the giant ants is located and then attacked during the day because the heat tends to drive the ants inside the nest. Cyanide gas grenades are hurled into the tunnel entrance of the giant nest. Graham, Peterson, and Patricia Medford, wearing gas masks and heavily armed, enter the nest and eventually reach the queen ant's chamber. (The inside of the nest is a wonderfully recreated version of a real ant nest, only large enough for humans to explore it.) The investigators discover that two new queen ants have already hatched and flown away from the nest.

The authorities begin a nationwide hunt for the escaped queen ants. One of them turns up on board a ship at sea. It apparently entered through an open hole while the ship was docked at port. The entire ship is infested with the giant ants, who are killed when naval gunfire sinks the ship.

The second queen ant has apparently landed in Los Angeles, since 40 tons of sugar have been reported stolen from a train boxcar. Eventually, the nest is determined to be located somewhere in the 700-mile long sewer system beneath Los Angeles.

The authorities finally announce the existence of the giant ants to the public and prepare for a major military operation to obliterate the monsters. A large number of patrols are sent into the sewer system, and the nest is located. Military personnel then converge on the nest and there is a furious battle against the ants until all of the giant worker ants are killed. When the troops enter the queen's chamber, they find that all the newly hatched queens are still in the nest. They are destroyed and humanity is saved. The picture ends with a thought-provoking question: if the first atomic bomb explosion produced the giant ants, what did the subsequent explosions create?

The movie has an excellent discussion about real ants and their nests, starting at 35 minutes into the film. Dr. Medford speculates about the cause of the mutations at 28 minutes into the film.

Fat Man and Little Boy
(United States) 1989, color, 126 minutes
Credits: Director, Roland Joffee; Producer, Tony Garnett.
Cast: Paul Newman (General Leslie Groves) and Dwight Schultz (J. Robert Oppenheimer).

Plot Summary:
This film describes the building of the first atomic bomb during World War II. It was constructed in the dirt and mud of Los Alamos, New Mexico, under the direction of General Leslie Groves, who had built the Pentagon. The endeavor was code-named the Manhattan Project. Among the scientists were refugees from Nazi-controlled Europe. They knew that they were in a race with Germany to develop a doomsday weapon that would decide the outcome of World War II. The practical problems in developing and testing the atomic bomb are described. Also the moral question of using the bomb is discussed by the scientists.

The film depicts the first atomic bomb blast (which was nearly postponed because of bad weather) and contrasts this success with the death of one of the scientists who accidentally received a lethal dose of radiation. (It does not, however, mention the concern of some of the real-life scientists that the blast might start a chain reaction that would destroy the entire planet.)

Technical discussions are found throughout the film. The accident that is fatal for one of the scientists occurs at 95 minutes into the film, and the first atomic bomb explosion occurs at 113 minutes into the film.

Chain Reaction
(United States) 1996, color, 106 minutes
Credits: Director, Andrew Davis; Producers, Arne Schmidt and Andrew Davis.
Cast: Keanu Reeves (Eddie Kosolivich), Morgan Freeman (Paul Shannon), and Rachel Weisz (Lily Sinclair).

Plot Summary:
This film describes the building of the first functioning hydrogen fusion energy generator at a fictitious University of Chicago Hydrogen Project with funding from a mysterious government agency. When the director of the project tells the agency representative, Paul Shannon, that he plans to reveal to the world the technical information needed to reproduce this technological breakthrough, he is killed by other agency operatives and the project self-destructs in a mini atomic explosion that devastates several square blocks of Chicago. One of the other researchers, Eddie Kosolivich, enters the laboratory just before its destruction and realizes that the blast was no accident. He and Lily Sinclair, another scientist working on the project, are framed for causing the disaster but they manage to elude the authorities as they piece together what happened. They are finally caught by agents of the government agency and brought to an underground facility where that agency is trying to reproduce their breakthrough.

They cause this government site to also self-destruct, but they escape the explosion with the help of Shannon. They then reveal the real culprits to the FBI. The world will have cheap energy after all!

The first 15 minutes of the film contains most of the science referenced in the film. There is a second segment depicting the attempt to reproduce the experiment in the underground facility which commences 70 minutes into the film.

FILMS RELATED TO CHAPTER 4: INDOOR AIR POLLUTION

The Incredible Shrinking Woman

(United States) 1981, color, 88 minutes
Credits: Director, Joel Schumacher; Producer, Hank Moonjean.
Cast: Lily Tomlin (Pat Kramer and Judith Beasley), Charles Grodin (Vance Kramer), and Ned Beatty (Dan Beame).

Plot Summary:
The film opens with an advertisement for a new artificial cheese product, one that squeezes from a bottle. A housewife, Pat Kramer, is depicted bringing home carloads of artificial items comprised of unknown chemical ingredients. Her home is filled with them and she is constantly exposed to the vapors emitted by these products and is occasionally drenched by some of these exotic chemicals.

Her husband works for an advertising agency that is paid to create a demand for these "wonders" of modern chemistry. For Pat, however, the exposure to one chemical too many has an unbelievable effect: she starts to shrink. Researchers at a mysterious institute examine her and conclude that she must have been especially susceptible to the chemicals she has experienced in her everyday life. When she returns home from the institute, the researchers begin a crash program to learn how to shrink the entire world's population, leaving only a select few as the "biggies" in charge of everything. The film is filled with humor, as when Pat, now only a few inches tall, is almost flushed down the garbage disposal and no one responds to her screams for help. Meanwhile, her husband is pressured by his boss, Dan Beame, not to publicly attribute his wife's condition to exposure to these artificial products in their home, since Beame's company promotes the very same products.

The mysterious research institute then kidnaps Mrs. Kramer in order to accelerate their research into shrinking the world's population. Pat manages to escape and warns the world just before she becomes so small that she vanishes in a pool of household chemicals (hair sprays, nail polishes, etc.). The ingestion of these new chemicals miraculously reverses her growth pattern and at the end of the film she has returned to her "normal" size but is still growing larger!

The film is a biting satire on society's preoccupation with things artificial.

Safe

(United States) 1995, color, 119 minutes
Credits: Director, Todd Haynes; Producers, Christine Vachon and Lauren Zalaznick.
Cast: Julianne Moore (Carol) and Xander Berkeley (Greg).

Plot Summary:
The film describes the deteriorating health of a housewife, Carol, due to worsening allergies to a variety of environmental factors. She and her husband, Greg, and his son by a previous marriage, occupy a huge home replete with all kinds of human-made contraptions and chemicals. At first Carol's physician downplays the seriousness of her symptoms. As her health worsens she is sent to a psychiatrist to deal with stress. When even that doesn't stop the progressive nature of her health problems, Carol goes to a secluded clinic for those suffering from environmental allergies. Those running the clinic believe that positive thinking can help each patient mend his or her immune system.

At first Carol improves, but at the end of the film she appears to be worse than ever. The film ends with her living in a sterile "safe house" looking at herself in a mirror and repeating "I love you." It is not clear how much of Carol's problems are in fact allergies and how much stem from her mental state.

The film is continuously filled with comments or examples of her condition and attempts to treat it.

FILMS AND TELEVISION PROGRAMS RELATED TO CHAPTER 5: THE GREENHOUSE EFFECT AND STRATOSPHERIC OZONE DEPLETION

Star Trek: The Next Generation—"A Matter of Time"
(United States) Episode 209, Airdate November 1991, 48 minutes
Credits: Director, Paul Lynch; Producer, Rick Berman.
Cast: Patrick Stewart (Captain Picard), Jonathan Frakes (Riker), Brent Spiner (Data), Gates McFadden (Dr. Crusher), LeVar Burton (La Forge), Michael Dorn (Worf), and Marina Sirtis (Troi).

Plot Summary:
The Enterprise is assisting the inhabitants of the planet Penthura IV, which had been struck by an asteroid. The crew is trying to reverse the nuclear-winter-like effects caused by the dust cloud hurled into the planet's atmosphere from the collision (which prevents 80% of the sunlight from reaching the planet's surface). They try to counteract this cooling of the planet by increasing the carbon dioxide in the atmosphere. This would increase the greenhouse effect, i.e. prevent more of the sun's energy that *does* penetrate the dust cloud from being reradiated back into space. The sources of the additional carbon dioxide are large underground pockets of the gas which the Enterprise's weapons free by blasting holes into the surface of the planet.

Unfortunately, the blasts triggers earthquakes and volcanic eruptions which spew even more dirt into the atmosphere, making matters worse. La Forge then proposes to clear away all of the dust from the atmosphere by using a carefully positioned burst of energy from the Enterprise, but this is risky since it may ignite the atmosphere of the planet, thereby killing all of its inhabitants. The government of the planet agrees to the experiment, which successfully disperses the dust cloud around Penthura IV.

The discussions of the greenhouse effect occur at 10 minutes and 18 minutes into the program.

Star Trek: The Next Generation—"When the Bough Breaks"
(United States) Episode 118, Airdate February 1988, 48 minutes
Credits: Director, Kim Manners; Producers, Rick Berman and Gene Roddenberry.
Cast: Patrick Stewart (Captain Picard), Jonathan Frakes (Riker), Brent Spiner (Data), Wil Wheaton, (Wesley Crusher), Gates McFadden (Dr. Crusher), LeVar Burton (La Forge), Michael Dorn (Worf), and Marina Sirtis (Troi).

Plot Summary:
In this episode from the first year of the *Star Trek: The Next Generation* television series, the Enterprise stumbles across a planet, Aldea, that is surrounded by a powerful shield which makes the planet invisible to outsiders. The starship has been led to Aldea by its inhabitants, who are unable to have children. The Aldeans kidnap seven of the children from the Enterprise. As Captain Picard negotiates with the Aldeans for the return of the children, Dr. Crusher learns that the Aldeans are dying from radiation poisoning, which has also made them sterile. The origin of the radiation poisoning is ultraviolet light that penetrates the planet's ozone layer, which has been badly

damaged by the shield around the planet. The supercomputer which maintains this shield is deactivated by the crew of the Enterprise and the Aldeans are promised help from the Federation to regain their health and restore their planet's ecology. Aldea will no longer be hidden from the rest of the universe by its shield, but its ozone layer will be rebuilt to block the ultraviolet light that had been harming its inhabitants.

The explanation of the Aldeans' health problems occurs at 39 minutes into the program.

Highlander II: The Quickening
(United States) 1991, color, 91 minutes
Credits: Director, Russell Mulcahy; Producers, Peter Davis and William Panzer.
Cast: Christopher Lambert (MacLeod), Sean Connery (Ramirez), Michael Ironside (the General), and Virginia Madsen (Louise).

Plot Summary:
As the film opens the year is 1999 and the ozone layer above the Earth continues to deteriorate. Millions are dying as a result. A shield is then established to block most of the sun's radiation from reaching the Earth, thanks to the scientific genius of MacLeod. The film then moves 25 years into the future, when society is rapidly deteriorating due to the effects of the shield —namely, that the temperature is a constant 97 degrees Fahrenheit and the humidity is also a constant 97% over the entire globe.

MacLeod now appears to be an infirm old man, but actually he is an alien from the planet Zeist who was banished to Earth 500 years ago by the planet's dictator, the General. Two assassins from Zeist are sent to kill MacLeod, and as a result of his successful battle with them MacLeod regains his immortality and youth. He befriends a young rebel who believes that the ozone layer has repaired itself, making the shield no longer necessary. MacLeod, assisted by Ramirez, another exile from Zeist, breaks into the control tower for the shield, disables it, and in the process kills the General, who has also journeyed to Earth to destroy MacLeod. The ozone layer has indeed repaired itself and the human race will live again in sunlight.

The description of the destruction of the ozone layer and the building of the shield both take place at the very beginning of the film. The director's cut version of the film describes the building of the shield 55 minutes into the film.

The Arrival
(United States) 1996, color, 115 minutes
Credits: Director, David Twohy; Producers, Thomas Smith and James Steele.
Cast: Charlie Sheen (Zane Zamensky), Ron Silver (Phil Gordon).

Plot Summary:
As the film opens, an observatory picks up a signal which appears to emanate from outer space indicating the existence of intelligent life elsewhere. One of the astronomers, Zane Zamensky, carries the news to his boss, Phil Gordon, but is immediately told that he is being laid off due to budget cuts.

Zamensky continues to monitor the skies for other signals with a home-made network of satellite dishes. His zeal leads his girlfriend to believe that he is mentally disturbed. His monitoring leads him to conclude that the signals are being answered by a station in Mexico. Zamensky travels to Mexico only to find the station destroyed. He then meets an American environmental scientist whose data indicates that the greenhouse effect is accelerating dramatically. They find a power station in Mexico which turns out to be a plant built and run by aliens to produce vast amounts of greenhouse gases. The aliens like it hot!

Zamensky avoids being killed by the aliens (one of whom turns out to be Gordon) and publicizes their diabolical plot to the world as the film ends.

The alien transmission is intercepted 3 minutes into the film. At 34 minutes into the film the greenhouse data is discussed. Finally, at 63 minutes, Zamensky enters the aliens' greenhouse gas-producing plant.

TELEVISION PROGRAM RELATED TO CHAPTER 6: ELECTROMAGNETIC WAVES

Star Trek: The Next Generation—"The Enemy"

(United States) Episode 155, Airdate November 1989, 48 minutes
Credits: Director, David Carson; Producer, Rick Berman.
Cast: Patrick Stewart (Captain Picard), Jonathan Frakes (Riker), Brent Spiner (Data), Wil Wheaton (Wesley Crusher), Gates McFadden (Dr. Crusher), LeVar Burton (La Forge), Michael Dorn (Worf), and Marina Sirtis (Troi).

Plot Summary:
In response to a distress call from a planet, Galorndon Core, located near the demilitarized border with the Romulan Empire, the Enterprise's away team finds a destroyed Romulan vessel and a badly injured Romulan soldier. La Forge, one of the members of the away team, falls into a cave and is thus not able to rendezvous with the others who beam back to the Enterprise. The planet is covered by severe electrical storms. Dr. Crusher finds that the injured Romulan has suffered damage to his synaptic connections which, she speculates, is caused by the strong magnetic fields (presumably associated with the electrical storms) found on the surface of the planet. She fears that La Forge is also in danger from these magnetic fields.

Meanwhile, LaForge manages to get out of the cave but is captured by Bochra, another Romulan soldier from the crashed ship. Efforts to rescue La Forge involve sending a neutrino beacon to the surface which can transmit a signal through the electrical storms to the Enterprise. La Forge persuades Bochra to help him find the neutrino beacon before they are both killed by the effects of the magnetic fields. The Romulan can hardly walk because of the damage to his nervous system from the magnetic fields, while La Forge's visor (which enables him to see) malfunctions because of these same magnetic fields. Working together, they find the beacon and signal the Enterprise, which beams them aboard.

Dr. Crusher's diagnosis of the Romulan's condition occurs at 10 minutes into the program. La Forge and Bochra discuss the effects of electromagnetic fields at 25 minutes into the program.